Automatic License Plate Detection and Recognition

Liu Yan   Chen Songlu   著

刘 艳   陈松路

Beijing

Metallurgical Industry Press

2025

Metallurgical Industry Press

39 Songzhuyuan North Alley, Dongcheng District, Beijing 100009, China

Copyright © Metallurgical Industry Press 2025. All rights reserved.

No part of this publication may be reproduced or transmitted in any form or by any means, electronic or mechanical, including photocopying, recording, or any information storage and retrieval system, without permission in writing from the copyright owner.

图书在版编目（CIP）数据

车牌自动检测与识别＝Automatic License Plate Detection and Recognition：英文／刘艳，陈松路著． 北京：冶金工业出版社，2025.1. —ISBN 978-7-5240-0084-6

Ⅰ. U491

中国国家版本馆 CIP 数据核字第 2025XF0687 号

Automatic License Plate Detection and Recognition

出版发行	冶金工业出版社	电　话	(010)64027926
地　　址	北京市东城区嵩祝院北巷 39 号	邮　编	100009
网　　址	www.mip1953.com	电子信箱	service@mip1953.com

责任编辑　戈　兰　美术编辑　彭子赫　版式设计　郑小利
责任校对　石　静　责任印制　范天娇

北京建宏印刷有限公司印刷
2025 年 1 月第 1 版，2025 年 1 月第 1 次印刷
710mm×1000mm 1/16；8.75 印张；220 千字；131 页
定价 99.00 元

投稿电话　(010)64027932　投稿信箱　tougao@cnmip.com.cn
营销中心电话　(010)64044283
冶金工业出版社天猫旗舰店　yjgycbs.tmall.com

（本书如有印装质量问题，本社营销中心负责退换）

Preface

The global proliferation of vehicles has presented substantial challenges in areas like road safety, traffic management, and law enforcement. Accurate and efficient license plate detection and recognition are critical to addressing these issues, forming the backbone of systems used in toll collection, parking management, and most importantly, law enforcement and traffic control. With the continuous advancement of machine learning and computer vision technologies, automatic license plate detection and recognition systems have become more sophisticated, capable of meeting the increasing demands for accuracy, speed, and adaptability.

This book delves deeply into the field of automatic license plate detection and recognition, aiming to provide a comprehensive guide to both researchers and practitioners. It offers a structured journey from foundational concepts to cutting-edge techniques, focusing on practical solutions to the various challenges posed by real-world applications. In simple scenarios, license plates are captured at short distances with minimal variation in angle and type. However, as demonstrated in this book, real-world conditions such as capturing plates from long distances, steep angles, and across diverse plate designs present far more complex detection and recognition problems. Moreover, a key contribution of this book is its focus on end-to-end approaches for solving the cascading problems often associated with traditional methods. By integrating detection and recognition into a unified network, this approach minimizes error accumulation and maximizes training efficiency. This book is designed to be an essential resource for those looking to stay at the forefront of automatic license plate detection and recognition technologies.

By blending theory, challenges, and real-world applications, this book aims to serve as a valuable resource for researchers, practitioners, and students involved in the field of computer vision and automatic license plate recognition. It is our hope that the insights and methodologies presented here will inspire further innovation and development in this dynamic area of research.

I would like to extend my heartfelt gratitude to all those who contributed to the development of this book. In particular, I wish to thank Liu Qi, Dai Songkang, Liu Yuanyuan, Mo Zhenlun, Sun Menglei, and Yang Siqi for their invaluable support and dedication throughout the process. Their insights, collaboration, and expertise have been instrumental in bringing this work to fruition. Special appreciation goes to the research community working in intelligent transportation systems, computer vision, and optical character recognition, whose tireless efforts continue to push the boundaries of what is possible in this domain. I am also deeply grateful to the institutions and organizations that have supported my work over the years. Their commitment to fostering innovation in the fields of machine learning and computer vision has been invaluable. Lastly, I would like to thank my family and friends for their unwavering encouragement and understanding throughout the writing process. Without their support, this book would not have been possible.

Author

Beijing, September 2024

Contents

Chapter 1 Introduction 1

  1.1 Motivation 1

  1.2 Outline of the book 2

Chapter 2 Simultaneous End-to-End Vehicle and License Plate Detection with Multi-Branch Attention Neural Network 6

  2.1 Problem formulation 6

  2.2 Problem formulation 8

  2.3 Experiments 13

Chapter 3 Scale-Invariant Multidirectional License Plate Detection with the Network Combining Indirect and Direct Branches 17

  3.1 Problem formulation 17

  3.2 Methodology 19

  3.3 Experiments 26

Chapter 4 Improving Small License Plate Detection with Bidirectional Vehicle-plate Relation 30

  4.1 Problem formulation 30

  4.2 Methodology 32

    4.2.1 Network architecture 32

    4.2.2 Training objective 33

  4.3 Experiments 35

    4.3.1 Datasets 35

    4.3.2 Valuation protocols 35

    4.3.3 Ablation study 36

    4.3.4 Comparative experiments 37

Chapter 5 Vertex Adjustment Loss for Multidirectional License Plate Detection and Recognition ... 41

5.1 Problem formulation ... 41
5.2 Methodology ... 42
5.3 Experiments ... 45
   5.3.1 Datasets ... 45
   5.3.2 Evaluation metrics ... 46
   5.3.3 Ablation study ... 46
   5.3.4 Comparative experiments ... 48
   5.3.5 License plate detection on CCPDv2 subsets ... 49
   5.3.6 License plate recognition on CCPDv2 ... 50
   5.3.7 Experiments on CLPD and RodoSol-ALPR ... 52
   5.3.8 Experiments on ICDAR2015 ... 53

Chapter 6 Fast Recognition for Multidirectional and Multi-type License Plates with 2D Spatial Attention ... 54

6.1 Problem formulation ... 54
6.2 Methodology ... 56
6.3 Experiments ... 59

Chapter 7 Improving Multi-type License Plate Recognition via Learning Globally and Contrastively ... 64

7.1 Problem formulation ... 64
7.2 Methodology ... 66
   7.2.1 Encoder ... 67
   7.2.2 Global modeling module ... 67
   7.2.3 Decoder ... 68
   7.2.4 Position-aware contrastive learning ... 70
   7.2.5 Prediction module ... 71
7.3 Experiments ... 72
   7.3.1 Datasets ... 72
   7.3.2 Experiments on benchmarks ... 74
   7.3.3 Ablation study ... 78
   7.3.4 Attention visualization ... 80

7.3.5	T-sne visualization	81
7.3.6	Discussion	82

Chapter 8   Towards Low-resource License Plate Recognition via Feature Shuffling 85

8.1   Problem formulation 85
8.2   Methodology 87
8.3   Experiments 91

Chapter 9   End-to-End Multi-line License Plate Recognition with Cascaded Perception 96

9.1   Problem formulation 96
9.2   Methodology 98
   9.2.1   Backbone 98
   9.2.2   License plate detection 99
   9.2.3   Recognition feature extraction 99
   9.2.4   License plate recognition 99
9.3   Experiments 101
   9.3.1   Datasets and evaluation metrics 101
   9.3.2   Ablation study 102
   9.3.3   Comparative experiments on multi-line license plates 104
   9.3.4   Comparative experiments on single-line license plates 106
   9.3.5   Cross-dataset experiments 108

Chapter 10   Multi-task Learning for License Plate Recognition in Unconstrained Scenarios 109

10.1   Problem formulation 109
10.2   Methodology 112
10.3   Experiments 114

References 122

Chapter 1   Introduction

1.1   Motivation

The proliferation of vehicles worldwide has brought about significant challenges in managing road safety, traffic control, and law enforcement. One of the most critical aspects of these efforts is the ability to accurately and efficiently detect and recognize license plates, which play a pivotal role in numerous applications, from toll collection and parking systems to law enforcement and traffic management. With the advent of advanced machine learning techniques and computer vision, automatic license plate detection and recognition systems have evolved to meet the growing demands for accuracy, speed, and adaptability.

This book delves into the core aspects of automatic license plate detection and recognition, aiming to provide a comprehensive understanding of the underlying principles, challenges, and state-of-the-art solutions. It is structured to offer a balanced approach, from foundational concepts to the most advanced techniques in the field.

In straightforward scenarios, license plates are captured from short distances, with limited variation in types and small angles. However, as illustrated in Fig. 1.1(a), in some cases, license plates are captured from a long distance, resulting in small, difficult-to-detect plates. Furthermore, as shown in Fig. 1.1(b), license plates may be captured at steep angles, often using mobile phones or dashcams, leading to multi-directional license plates. These factors significantly complicate detection. Additionally, as depicted in Fig. 1.1(c), license plates around the world vary greatly in type, featuring different

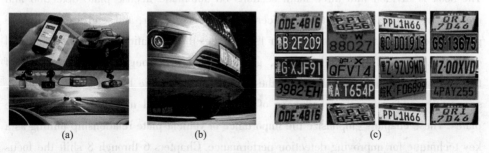

Fig. 1.1   License plates captured with different distances, angles, and types

fonts, backgrounds, and character layouts, further increasing the challenge of accurate recognition.

Furthermore, as depicted in Fig. 1.2 (a), license plate recognition is typically achieved by cascading separate detectors and recognizers, which involves extracting recognition features from the original image. However, this approach can lead to error accumulation and reduce training efficiency due to the involvement of multiple models. To solve these problems, researchers have proposed end-to-end methods. As depicted in Fig. 1.2(b), these methods extract recognition features from deep layers of the CNN backbone. Unfortunately, these deep layers undergo multiple down-sampling operations, losing crucial license plate information. As a result, this process hinders accurate license plate recognition. To solve the above problems, as depicted in Fig. 1.2 (c), we propose to construct an end-to-end LPR network that optimizes detection and recognition jointly. In this approach, both the detection and recognition networks share the same CNN backbone. Specifically, the detection network utilizes deep-layer features from the backbone due to their strong semantics and large receptive fields. On the other hand, the recognition network leverages features from shallow layers of the backbone, which better preserves crucial license plate information, such as character stroke, thereby enhancing the subsequent recognition.

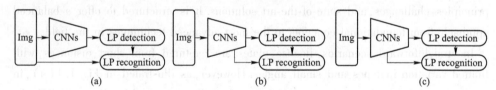

Fig. 1.2 Three different sources for extracting recognition features

1.2 Outline of the book

The book is divided into three main sections on automatic license plate detection and recognition: license plate detection, license plate recognition, and end-to-end license plate detection and recognition. Chapter 1 introduces the motivation behind this work and outlines the structure of the book, guiding readers through the progression of topics. Chapters 2 through 5 focus on the nuances of automatic license plate detection, specifically addressing the challenges of detecting small and multi-directional license plates. These chapters emphasize the importance of vehicle-plate relationship mining as a key technique for improving detection performance. Chapters 6 through 8 shift the focus to automatic license plate recognition, discussing the critical aspects of fast and accurate recognition. The chapters explore how to handle different types of plates and environments where resources, such as high-quality imagery or computational power, are limited. In Chapters 9 through 10, the book presents the cutting-edge approach of end-to-end license plate detection and recognition. These chapters explore the integration of detection and recognition into a unified framework, offering insights into how such systems can achieve higher levels of efficiency and effectiveness. The chapter overview is illustrated in Table 1.1.

Table 1.1 Chapter overview

Section	Chapter No.	Problem Formulation	Methodology
License plate detection	2	Simultaneous vehicle and license plate detection	Multi-branch attention neural network
	3	Small and multi-directional license plates detection	Combining direct detection and vehicle-plate relation; Vertex regression
	4	Small license plates detection	Bidirectional vehicle-plate relation
	5	Multi-directional license plate detection	Vertex adjustment
License plate recognition	6	Multi-type license plate recognition	Spatial attention
	7	Multi-type license plate recognition	Learning globally and contrastively
	8	Low-resource license plate recognition	Feature shuffling
End-to-end license plate detection and recognition	9	End-to-end multi-line license plate detection and recognition	Cascaded perception
	10	End-to-end multi-type license plate detection and recognition	Multi-task learning

Chapter 2 discusses the challenge of efficiently detecting both vehicles and license plates simultaneously in most cases. A single network may cause the vehicle to interfere with license plate detection due to the inclusion relationship between the two. To address this, we propose a deep neural network that detects both vehicles and license plates in a given image. This network uses two separate branches with distinct convolutional layers for vehicle and license plate detection, respectively.

Chapter 3 introduces a network combining both indirect and direct branches for license plate detection in challenging environments. The indirect branch detects small license plates with high precision in a coarse-to-fine manner using vehicle-plate relationships. Meanwhile, the direct branch detects the license plate directly from the input image, reducing false negatives caused by missed vehicle detections in the indirect branch.

Chapter 4 notes that pre-detecting the vehicle can improve the detection of small license plates. However, this method relies solely on a one-way relationship, where the presence of a vehicle enhances license plate detection, which may lead to error accumulation if the vehicle is not detected. To solve this, we propose a unified network that simultaneously detects both vehicles and license plates while establishing a bidirectional relationship between them. This approach improves small license plate detection and reduces error accumulation when the vehicle is missed.

Chapter 5 discusses the rigid nature of license plates, which only undergo rigid deformation in images. Many researchers address multidirectional license plate detection by regressing the vertices of the plate. However, treating each vertex as an independent task ignores the relationships between vertices, leading to inconsistencies in regression accuracy. We propose using vertex relationships to improve predictions for multidirectional license plate detection. Specifically, after regressing the vertices, we align the minimal rectangle they form with the ground truth, adjusting any deviation. This adjustment occurs only during training, adding no extra network parameters.

Chapter 6 addresses the challenge of recognizing multidirectional and two-line license plates. When a license plate image is transformed directly into a one-dimensional feature sequence, the features of adjacent characters may become blurred. To resolve this, we propose a two-dimensional spatial attention module that recognizes license plates from a two-dimensional perspective.

Chapter 7 highlights the difficulties of multi-type license plate recognition due to varied character layouts and fonts. Two main issues arise: recognition models often misinterpret character locations because of diverse layouts, and similar glyphs in different fonts may lead to character misidentification. To overcome these problems, we propose two plug-and-play modules within an attention-based framework to improve multi-type license plate recognition.

Chapter 8 focuses on the challenge of recognizing characters in small-scale, low-resource license plate datasets, which tend to exhibit long-tailed distributions in certain character classes due to limited character permutations. Existing methods often prioritize high-frequency (head) classes, leaving low-frequency (tail) classes underrepresented. To solve this, we propose feature shuffling to balance the distribution across character classes, improving recognition for tail classes. Additionally, we introduce global perception to better understand overall character layouts, enhancing feature shuffling.

Chapter 9 highlights the challenge of recognizing multi-line license plates due to their irregular layouts. Previous methods have struggled to recognize them effectively. In this work, we propose an end-to-end multi-line license plate recognition network that cascades global type perception with parallel character perception, enhancing both recognition accuracy and speed.

Chapter 10 presents an end-to-end method for license plate detection and recognition using multi-task learning. We introduce two parallel branches: one detects the horizontal bounding box, and the other identifies the four corners of the license plate, enabling multidirectional license plate detection. These outputs are combined to improve recognition accuracy. Additionally, we extract global features to understand character layouts and use reading order to spatially focus on characters, improving multi-line license plate recognition. Finally, by using the same backbone, with the detection branch based on deep layers and the recognition branch based on shallow layers, we create an end-to-end network for efficient detection and recognition.

Please see colorful images

Chapter 2 Simultaneous End-to-End Vehicle and License Plate Detection with Multi-Branch Attention Neural Network

Vehicle and license plate detection plays an important role in intelligent transportation systems and is still a challenging task in real applications, such as on-road scenarios. Recently, Convolutional Neural Network (CNN)-based detectors achieve the state-of-the-art performance. However, it is difficult to efficiently detect the vehicle and license plate simultaneously in most cases. With a single network, the vehicle can affect the detection of the license plate due to the inclusion relation. To solve this problem, we propose an end-to-end deep neural network for detecting the vehicle and the license plate simultaneously in a given image, where two separate branches with different convolutional layers are designed for vehicle detection and license plate detection, respectively. In consideration of the license plate's small size and fairly obvious features as well as the vehicle's various size and rather complex features, the license plates are detected with low-level features and the vehicles are localized with multi-level features in corresponding convolutional layers. Moreover, a task-specific anchor design strategy is employed to obtain better predictions. Besides, the attention mechanisms and feature-fusion strategies are utilized to improve the detection performance of small-scale objects. A variety of experiments on real datasets and public datasets verify that our proposed method has fairly high accuracy and efficiency.

2.1 Problem formulation

Automatic vehicle and license plate detection are important in intelligent transportation systems. A variety of methods have been proposed in the literature. However, license plate detection is still considered a challenging task in real applications because of the small size of captured license plates, illumination variations in the scene, and viewpoint changes of cameras. Similarly, the problem of vehicle detection in real scenes is also unsolved because of vehicle size changes, vehicle pose variations, and complex scene backgrounds.

The first row is for SSD while the second row is for ours. With SSD, some small-scale vehicles and license plates are not detected, and the confidence of detected license plates is at a low level. All recognizable license plates are manually blurred to protect privacy.

There have been several powerful object detection methods, e. g. , SSD, YOLO, and Faster R-CNN. However, we find it difficult to detect the vehicle and license plate simultaneously using these prestigious frameworks. As seen in Fig. 2. 1, some distinct license plates are unexpectedly failed to be detected, and the confidence of detected ones is also at a low level. In deep neural networks, the vehicle and the license plate generally share the same head networks and anchor boxes. Thus, license plate detection is easily affected by the vehicle because of their inclusion relation.

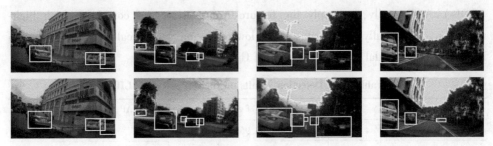

Fig. 2. 1　Examples from SSD and our method

Table 2. 1 demonstrates the detection results on the test set of VALID with SSD300 when training vehicle and license plate separately in two independent networks or simultaneously in one network. We can see that the AP of the license plate drops more than 14% when training vehicle and license plate together. Meanwhile, the AP of the vehicle is not affected. To illustrate the vehicle's effects on the license plate, all license plates are randomly distributed to other places to decouple the relationship between the vehicle and the license plate, making sure the distributed license plates are not overlapped with vehicles and each other. From the first two rows of Table 2. 2, we can see that removing the license plate does not affect the vehicle. However, the performance of the license plate improves a lot after separating it from the vehicle, almost having the same performance as training license plate alone in Table 2. 1. To solve this problem, we propose an end-to-end multi-branch attention neural network for simultaneously detecting the vehicle and the license plate in a given image, where two separate branches with different convolutional layers are stemmed from the backbone network to detect the vehicle and the license plate respectively. In general, the low-level features of CNNs have high resolution with weak semantics and are important to small object detection. Meanwhile, high-level features are semantically strong but with low resolution, and these features have better feature representation of large objects. Intuitively, the license plates usually have a relatively small size and fairly simple features, while the vehicles have various scales and rather complex features. Accordingly, we assign low-level features for license plate detection and multi-level features for vehicle localization. Moreover, a task-specific anchor design strategy is also applied for vehicle and license plate detection. For both two kinds of objects, we select better anchor priors instead of hand-picked ones based on the clustering method, which can make it easier to learn better predictions. Besides, as shown in Fig. 2.1, small objects are always failed to be detected, especially for the license plates and vehicles in the distance. We add the spatial attention mechanisms into each branch to facilitate focusing on the regions of interest (ROIs). Additionally, we apply the feature-fusion strategy of combining both high-resolution, semantically weak features and low-resolution, semantically strong features to leverage the pyramidal feature hierarchy of CNNs.

Table 2.1 Detection results on the test set of VALID

Method(Classes)	$AP_{0.5}$ of vehicle	$AP_{0.5}$ of license plate
SSD(Vehicle)	83.96%	—
SSD(License plate)	—	85.90%
SSD(Vehicle+license plate)	83.77%	71.86%

Table 2.2 Detection results on the processed test set of VALID with SSD300 and two branches

Method(Classes)	$AP_{0.5}$ of vehicle	$AP_{0.5}$ of license plate
SSD(Vehicle w/o LP)	84.28%	—
SSD(Vehicle w/o LP + Dis LP)	83.61%	85.26%
TB(Vehicle w/o LP + Dis LP)	84.23%	84.11%

2.2 Problem formulation

We propose an end-to-end multi-branch attention neural network to detect vehicle and license plate simultaneously, where two separate branches with different convolutional layers are designed for vehicle detection and license plate detection respectively. The license plates are detected with low-level features and the vehicles are localized with multi-level features. Moreover, a task-specific anchor design strategy is applied for better object predictions. Besides, we employ the attention mechanisms and feature-fusion strategies to improve the recall of small-scale cases. The overall network architecture is demonstrated in Fig. 2.2.

2.2 Problem formulation

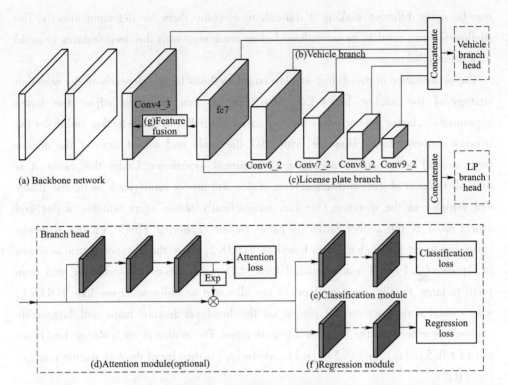

Fig. 2.2 Network architecture

The backbone network is inherited from the popular VGG-16, keeping convolutional layers from Conv1-1 to Conv5-3. The last two fully-connected layers (fc6, fc7) are converted into convolutional layers and extra layers from Conv6-2 to Conv9-2 are also added for semantically stronger feature extraction, which is the same with SSD for a fair comparison. It is difficult to detect the vehicle and license plate simultaneously using the same head networks (classifier and regressor) and anchor boxes because of the inclusion relation. The vehicle can easily affect the detection of the license plate when these two objects act on the same anchor boxes through a shared classifier and regressor. Thus, it is necessary to decouple the relationships between the two objects, and vehicle and license plate detection can be separated into different branches. Each branch holds its head network for object classification and bounding box regression respectively. Moreover, low-level features can be useful for license plate detection due to the small size and fairly obvious features, and multi-level features can be used for vehicle localization due to the various scales and rather complex features. As illustrated in Fig. 2.2, we assign several relatively shallow layers to the license plate detection branch and allocate dispersed layers to the vehicle detection branch. Moreover, the scale of features in different layers may be quite different, making it difficult to combine them for detection directly. The shallow features need to be normalized before combining with the deep features to avoid parameter imbalance.

The performance of the sliding-window based methods largely depends on the selection strategy of the anchor boxes. Even the network can learn to adjust the boxes appropriately, better anchor priors make it easier to predict better detection and make the network converge faster. However, with SSD, the scale and aspect ratio of the anchor boxes are all set manually according to empirical experience. Under this mode, it is difficult to cover objects with uncommon scales and aspect ratios, such as license plates and vehicles in the distance. One can automatically obtain more suitable scales and aspect ratios of the anchor boxes by using anchor clustering. Firstly, we run K-means clustering to get the anchor priors based on YOLOv2, where the distance metric is shown in Equation(2.1)(GT means ground truth). Then, all priors are sorted by area from small to large. Finally, the sorted priors are allocated to different layers like YOLOv3, where small-scale anchors are placed on the low-level feature maps and large-scale anchors are placed on the high-level feature maps. The center of each anchor box is set to $((i+0.5)/|s_k|, (j+0.5)/|s_k|)$, where $|s_k|$ is the size of the k-th feature map, i, $j \in [0, |s_k|]$.

$$d(GT_{box}, \text{centroid}) = 1 - \text{IoU}(GT_{box}, \text{centroid}) \qquad (2.1)$$

Moreover, the average IoU can be calculated with the closest centroid without considering the spatial position of the anchor boxes. The average IoU is computed under the ideal conditions. However, anchor boxes of SSD-based methods are scattered in a sparse way, where the average IoU should be calculated with the spatial anchor boxes and we call it spatial IoU.

According to Fig. 2.1, in real scenes, it is challenging to detect small-scale objects, especially for the license plates and vehicles in the distance. The spatial attention mechanism can highlight the foreground information and keep the context information. We add a segmentation-like mask before the classification and regression module. Only the two shallowest layers in each detection branch adopt the attention module because deeper layers have large receptive fields and easily bring in noises. For both branches, the attention module helps to highlight features of the foreground regions and diminish the background regions. The attention supervision information is simply obtained by filling the ground truth and the attention loss is simply pixel-wise sigmoid cross-entropy between the filled ground truth and the predicted mask. Finally, the attention maps are fed into exponential operation and then have dot product with the feature maps. Fig. 2.3 shows Attention mask. Each row shows one original image resized to 300×300, covered with the attention mask. The first two columns demonstrate the attention masks of the vehicle, and the last two columns demonstrate the attention masks of the license plate. Fig. 2.3 demonstrates two examples of the predicted attention mask. As can be seen, it tends to focus on the center of objects. Moreover, the attention mask of high-level feature maps can cover more RoIs because of larger respective fields. In each detection branch, the attention module helps to enhance the foreground information, which is favorable to small-scale objects.

Fig. 2.3 Attention mask

Furthermore, fusing high-level features with low-level features can enhance the semantic representation. To further reduce computational complexity, we simply utilize the feature fusion strategy in the two shallowest layers of each detection branch, as illustrated in Fig. 2.4. To achieve the speed-accuracy tradeoff, we adopt FPN as the feature fusion module. The upper layer is firstly up-sampled by a factor of 2 using nearest-neighbor interpolation, and then undergoes a 1×1 convolutional layer to reduce channel dimensions. Considering the different scales between different layers, the upper layer needs to be rectified by ReLU and then merged with the normalized low-level features by element-wise addition. Fig. 2.4 illustrates the building block of the merging operation between lateral connection and top-down pathway.

The optimization function is composed of three parts. For the classification regression module, we adopt the same loss function as SSD. Let c be the confidence, l be the predicted box, g be the ground truth box, N be the number of matched anchor boxes. For

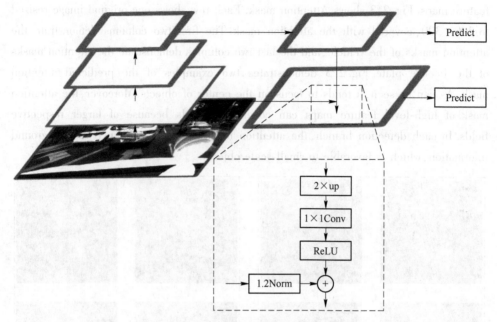

Fig. 2.4　Feature fusion building block

the attention module, we calculate the pixel-wise sigmoid cross-entropy between the generated attention mask and the ground truth. Let K be the index of all used pyramidal features, m_a^k be the attention mask generated per level, m_g^k be the ground truth of the attention mask. α and β are the weighting parameters to balance these terms. We utilize two separate optimization functions for the vehicle detection branch and the license plate detection branch respectively. These two branches undergo the separate back-propagation process, and γ is the weighting factor to adjust the two branches. The loss function is defined as Equations (2.2) and (2.3), and we simply set $\alpha = \gamma = 1$ and $\beta = 3$. We adopt the smooth L_1 loss and the softmax loss for regression and classification respectively and employ the pixel-wise sigmoid cross-entropy for attention loss. Let $K_t = 2$ be the class number when training vehicle and license plate together with shared head layers, $C_t = \{\text{vehicle}, \text{LP}, \text{background}\}$ be the classes. With SSD, the confidence of the n-th object is calculated by Equation (2.4).

$$L(x,c,l,g,m) = \frac{1}{N}(L_{\text{conf}}(x,c) + \alpha L_{\text{loc}}(x,l,g)) + \beta \sum_{k \in K} L_a(x, m_a^k, m_g^k) \quad (2.2)$$

$$L_{\text{total}}(x,c,l,g,m) = L_{\text{vehicle}}(x,c,l,g,m) + \gamma L_{\text{LP}}(x,c,l,g,m) \quad (2.3)$$

$$\text{con} f_t^n = \frac{e^{C_t^n}}{\sum_{m=1}^{K_t+1} e^{C_t^n}} \quad (2.4)$$

Let $K_V = 1$ be the class number of the vehicle detection branch, $C_V = \{\text{vehicle}, \text{background}\}$ be the classes. Let $K_{LP} = 1$ be the class number of the license plate detection branch, $C_{LP} = \{\text{LP}, \text{background}\}$ be the classes. With our method, the confidence of the n-th object is calculated separately for each detection branch by Equations (2.5) and (2.6).

$$\text{con} f_V^n = \frac{e^{C_V^n}}{\sum_{m=1}^{K_V+1} e^{C_V^n}} \quad (2.5)$$

$$\text{con} f_{LP}^n = \frac{e^{C_{LP}^n}}{\sum_{m=1}^{K_{LP}+1} e^{C_{LP}^n}} \quad (2.6)$$

As for the regression module, it predicts the offsets of position and scale to the anchor boxes. Suppose a anchor box $d = (x_d, y_d, w_d, h_d)$ and the predicted values $(\Delta x, \Delta y, \Delta w, \Delta h)$, the box $b = (x, y, w, h)$ can be obtained as follows.

$$x = x_d + w_d \Delta x \quad (2.7)$$
$$y = y_d + w_d \Delta y \quad (2.8)$$
$$w = w_d \exp(\Delta w) \quad (2.9)$$
$$h = h_d \exp(\Delta h) \quad (2.10)$$

2.3 Experiments

We adopt VGG-16 as the base model, which is pre-trained on the ILSVRC CLS-LOC dataset. The baseline network follows SSD300. All the training images are augmented with random crop and distortion, etc., following the same scheme as SSD. Our model is trained with 300×300 images using Adam for 40k iterations. The momentum parameters are set to $\beta_1 = 0.9$ and $\beta_2 = 0.999$. Learning rate, weight decay, and batch size are set to $10^{-4}, 5 \times 10^{-4}, 32$ respectively. All the experiments are carried on a PC with 4 NVIDIA TITAN Xp GPUs. We employ two auto-mobile data recorders to collect videos on the road of a Chinese city with the resolution of 720 × 1280. For simplicity, we name our dataset VALID (Vehicle And LIcense plate Dataset). A total of 887 images are collected and carefully annotated. 78 images from one recorder are used as the test set. The rest 809 images from another recorder are randomly divided into the training set and the validation set by 7 : 3. DETROIT We re-annotate a subset from Open Image Dataset (OID) V4, which contains "Car" and "Vehicle registration plate". For simplicity, we call it DETROIT (DatasET fRom Open Image daTaset). The images of DETROIT are obtained from the Internet, and the size and aspect ratio varies greatly. 386 images from the OID validation set are used as the test set. 1113 images from the OID test set are randomly divided into the training set and the validation set by 7 : 3. DOC We combine vehicle position from Cars and license plate position to obtain DOC (Dataset frOm Cars). A total of 105 images are obtained. 70% are randomly selected as the training-validation set, and the rest 30% is used as the test set. The images of DOC are also obtained from the Internet, and the size and aspect ratio varies greatly. Due to the inclusion relation, the classifier tends to classify candidate anchor boxes as vehicles. Furthermore, with SSD, the regressor is also class-agnostic, which can make the regression parameters unstable because they are influenced by the vehicle and the license plate simultaneously. Based on this, we first separate the classifier for the vehicle and the license plate and then also separate the regressor for two objects into two independent detection branches.

Table 2.3 demonstrates the detection results of sharing classifier and regressor, only sharing regressor as well as separating both classifier and regressor. It can be seen that the performance of the license plate improves a lot after separating the classifier, which proves that the vehicle can affect license plate detection. Besides, the performance of the vehicle drops a bit after separating the classifier because we intuitively think the license plate can influence more on the regressor. Moreover, the performance of the vehicle and the license plate obtains further improvement after separating the regressor because of independent detection branches. In consideration of the license plate's small size and fairly obvious features as well as the vehicle's various size and rather complex features, the license plates are detected with low-level features and the vehicles are localized with multi-level features.

Table 2.3 Ablation study on the test set of VALID

Method(Classes)	Classifier separation	Two branches	Anchor clustering	Attention	Feature fusion	$AP_{0.5}$ of vehicle	$AP_{0.5}$ of LP
SSD(Vehicle)						83.96%	—
SSD(LP)						—	85.90%
SSD(Vehicle+LP)						83.77%	71.86%
Ours(Vehicle+LP)	√					82.60%	82.35%
	√	√				83.78%	85.48%
	√	√	√			85.69%	86.21%
	√	√	√	√		86.37%	87.69%
	√	√	√	√	√	**86.84%**	**88.28%**

We apply K-means clustering on the training set of VALID to generate anchor priors of the vehicle and the license plate respectively. With SSD, altogether 30 kinds of anchor boxes are adopted, denoted as $A_{SSD} = \{4,6,6,6,4,4\}$ for 6 head layers respectively. We select 12 cluster centroids for the vehicle and 10 cluster centroids for the license plate, denoted as $A_{Vehicle} = \{4,6,1,1,1,1\}$ and $A_{LP} = \{4,6\}$ respectively. For each branch, we assign a close number of anchor boxes with SSD. For the vehicle detection branch, the last 4 head layers share 2 cluster centroids. The average IoU and spatial IoU of the license plate are all at a low level using the anchor design strategy of SSD. Our anchor clustering method achieves higher average IoU and spatial IoU with fewer cluster centroids and anchor numbers, especially for the license plate. Furthermore, due to better matching with the ground truths, our strategy makes the network converge faster.

Both the attention mechanism and feature-fusion strategy improve the detection performance for both two objects. The spatial attention mechanism can highlight foreground information for better detection. The predicted attention mask is added before the classification and regression module, where the attention maps are fed into exponential operation and then have dot product with the feature maps. In this way, the regions of the vehicle and license plate are enhanced while the background is kept. Considering that deeper layers have larger receptive fields, attention on these layers may bring in extra noises. Only the bottom two layers of each branch are employed with attention. Furthermore, we also simply apply the feature fusion strategy between the bottom two layers of each branch.

For VALID, DETROIT and DOC, we compare Faster R-CNN, YOLO, YOLOv2, YOLOv3 and SSD with our proposed method. The backbone of Faster R-CNN and SSD is set to VGG-16, while the backbone of YOLO (v1-v3) remains unchanged. For our method, the experiment settings follow the settings of VALID, including the anchor clustering centroids. As shown in Table 2.4, whether training vehicle and license plate separately or together, our method obtains the best performance for both three datasets.

Table 2.4 Detection results on VALID, DETROIT, and DOC

Classes	Method	VALID		DETROIT		DOC	
		Vehicle	LP	Vehicle	LP	Vehicle	LP
Vehicle	Faster R-CNN 300	71.15%	—	68.33%	—	**100%**	—
	YOLO 320	69.25%	—	63.11%	—	57.32%	—
	YOLOv2 320	76.73%	—	71.89%	—	96.88%	—
	Fast YOLOv2 448	75.02%	—	65.11%	—	96.88%	—
	Fast YOLOv3 320	79.42%	—	69.06%	—	**100%**	—
	SSD 300	83.96%	—	71.37%	—	**100%**	—
	Ours 300	**86.13%**	—	**71.92%**	—	**100%**	—

Continued Table 2.4

Classes	Method	VALID		DETROIT		DOC	
		Vehicle	LP	Vehicle	LP	Vehicle	LP
LP	Faster R-CNN 300	—	54.95%	—	62.25%	—	59.56%
	YOLO 320	—	66.75%	—	64.82%	—	49.18%
	YOLOv2 320	—	80.75%	—	73.95%	—	96.68%
	Fast YOLOv2 448	—	79.13%	—	67.10%	—	89.93%
	Fast YOLOv3 320	—	79.69%	—	69.30%	—	90.62%
	SSD 300	—	85.90%	—	75.96%	—	96.78%
	Ours 300	—	**88.73%**	—	**79.73%**	—	**97.07%**
Vehicle+LP	Faster R-CNN 300	72.09%	32.02%	67.03%	37.59%	**100%**	9.63%
	YOLO 320	70.42%	61.07%	63.72%	62.73%	56.54%	31.53%
	YOLOv2 320	76.35%	68.31%	70.01%	68.80%	96.88%	93.65%
	Fast YOLOv2 448	74.32%	60.15%	64.51%	62.75%	**100%**	82.51%
	Fast YOLOv3 320	79.77%	79.09%	68.71%	69.20%	**100%**	90.62%
	SSD 300	83.77%	71.86%	71.51%	71.52%	**100%**	95.46%
	Ours 300	**86.84%**	**88.28%**	**72.49%**	**79.29%**	**100%**	**96.88%**

Moreover, with our method, the performance of the license plate improves greatly for both three datasets. For other methods, except YOLOv3, they all have similar phenomena with SSD 300 when training vehicle and license plate together. License plate detection is largely affected by the vehicle, while the vehicle is less affected. YOLOv3 divides the input images into many grids and each grid is responsible for detecting the object, so license plate detection is almost unaffected. For Faster R-CNN, the performance of the license plate drops dramatically for both three datasets, because Faster R-CNN is a two-stage network and license plate detection can be affected by the vehicle in both two stages.

Please see colorful images

Chapter 3 Scale-Invariant Multidirectional License Plate Detection with the Network Combining Indirect and Direct Branches

As the license plate is multiscale and multidirectional in the natural scene image, its detection is challenging in many applications. In this work, a novel network that combines indirect and direct branches is proposed for license plate detection in the wild. The indirect detection branch performs small-sized vehicle plate detection with high precision in a coarse-to-fine scheme using vehicle-plate relationships. The direct detection branch detects the license plate directly in the input image, reducing false negatives in the indirect detection branch due to the miss of vehicles'detection. We propose a universal multidirectional license plate refinement method by localizing the four corners of the license plate. Finally, we construct an end-to-end trainable network for license plate detection by combining these two branches via post-processing operations. The network can effectively detect the small-sized license plate and localize the multidirectional license plate in real applications. To our knowledge, the proposed method is the first one that combines indirect and direct methods into an end-to-end network for license plate detection. Extensive experiments verify that our method outperforms the indirect methods and direct methods significantly.

3.1 Problem formulation

License plate detection (LPD) plays an essential role in many practical applications, including electronic toll collection, traffic surveillance, and enforcement. When the image acquisition conditions (shooting distance and angle) are restricted, such as the parking toll, the LPD task is almost completely solved. However, if the image is captured in the wild, it remains challenging due to various sizes, orientations, and backgrounds. Fig. 3.1 illustrates some license plate (LP) examples in real scenarios.

Recent LPD methods can be roughly divided into direct and indirect ways. Direct methods directly localize the license plate in the input image with handcrafted features, deep learning features, or license plate recognition system. How-ever, detecting small-sized license plates is challenging since they only occupy a relatively small area in the whole image. Indirect methods detect the license plate using the vehicle's proposal or the vehicle head region. The vehicle head region is manually defined as the smallest region enclosing the headlights and tires. The indirect methods can reduce the detection area and background noises, which is favorable to small-sized license plate detection. However, when the vehicle fails to be detected due to severe occlusion or nonuniform illumination, it will fail to localize the license plate.

Fig. 3.1 License plates in real scenarios

To overcome these problems, we propose a novel network composed of an indirect branch and a direct branch. The indirect detection branch can approximately localize the license plate based on the spatial relationships between the license plate and the vehicle. Then it can refine the license plate in the local region. This way, it can significantly reduce the detection area and mitigate the adverse effects of the background noises, which is favorable to small-sized license plate detection. The direct detection branch can reduce false negatives in the indirect detection branch due to the miss of vehicles'detection. We combine the indirect and direct branches to construct an end-to-end trainable network for license plate detection. The detection results of two branches are merged by post-processing operations, such as non-maximum suppression (NMS). Extensive experiments show that our method outperforms both the direct approach and the indirect approach.

Moreover, many methods do not consider the orientation of the license plate, which is only applicable to specific scenarios, such as parking charges and vehicle access/exit management. When it comes to more complex scenarios, such as road scenes, if we regard the tilted license plate as the horizontal direction, it may cause errors in the subsequent license plate recognition. Although Dong et al. propose to detect the multidirectional license plate, these methods are very complicated due to adopting multiple separate models.

We propose to detect the multidirectional license plate by localizing the four corners of the license plate to reduce complexity. It can be easily implemented by integrating the corner prediction module into the two branches mentioned above, with no extra models. In this way, the whole detection network is still in an end-to-end trainable manner, as shown in our open-source codes.

3.2 Methodology

We propose a novel network for license plate detection, which can effectively detect the small-sized license plate and accurately localize the multidirectional license plate in real applications. The indirect detection branch can precisely detect the small-sized license plate. The direct detection branch can reduce the false-negative license plate in the indirect detection branch due to incorrectly detected vehicles. The whole network is constructed in an end-to-end trainable manner. The detection results of these two detection branches are merged by post-processing operations, such as NMS.

The overall architecture is illustrated in Fig. 3.2. The network is constructed with two detection branches, i.e., indirect detection branch and direct detection branch. In the indirect detection branch, the approximate location and size of the license plate are predicted at the ALPD stage, where the center of the license plate (green circle) is obtained based on the offset (purple arrow) between the center of the license plate and the vehicle (orange circle). Moreover, the probability of the vehicle containing a license plate (red number) is predicted simultaneously. At the LREA stage, the local region of LP is obtained by expanding the LP region, and all the expanded LP regions (green dashed rectangle) are resized and aggregated into feature patches via differentiable region of interest (RoI) warping for batch operation. At the MLPR stage, the quadrilateral (red circle) and horizontal (green rectangle) bounding boxes of the license plate are detected simultaneously in the local region of LP. In the direct detection branch, the license plate is directly detected in the input image at the DLPD stage. The DLPD and ALPD modules share the same backbone network but different detection head networks. Finally, the detection results of two branches are merged by post-processing operations, such as NMS. The network can be trained in an end-to-end manner, where the red arrows denote the backpropagation gradients.

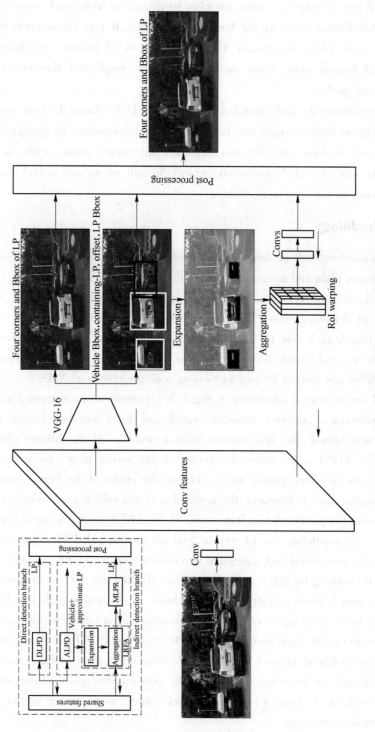

Fig. 3.2 A thumbnail of the overall architecture

The indirect detection branch predicts the approximate location of the license plate utilizing spatial vehicle-plate relationships firstly, then estimates the local region by expanding the LP region followed by an aggregation operation, and refines the quadrilateral and horizontal bounding boxes of the license plate in the local region finally. This multi-level design enables the model to focus on the potential location of the license plate and reduce the disturbing background noises.

At this stage, the approximate location of the license plate is estimated according to the vehicle-plate relation. At first, the vehicle is detected, so the center of the vehicle is determined. After that, the location of the license plate is obtained based on the offset between the center of the license plate and the vehicle. Meanwhile, the size of the license plate is directly predicted in the input image. According to the center and size, the license plate is approximately detected. In addition, the probability of the vehicle containing a license plate is predicted simultaneously. As shown in Fig. 3.2, the location and size of the license plate are not accurate in general cases because the license plate only occupies a relatively small area in the large input image.

The ALPD module is based on SSD for multi-task learning, which is the same as SSD512 except for the training objective. The training objective of the ALPD module is defined as Equation (3.1), including five losses: vehicle classification loss $L_{cls}(c)$, vehicle regression loss $L_{reg}(p,g)$.

$$L_1(c,p,g) = \frac{1}{N_v}[L_{cls}(c) + L_{reg}(p,g) + L_{off}(p,g) + L_{size}(p,g) + L_{con\_lp}(p,g)]$$
(3.1)

where, N_v is the number of matched anchor boxes with the ground-truth vehicles, c is the vehicle presence confidence, p is the predicted parameters, and g is the ground-truth parameters.

The training objective of vehicle detection is derived from SSD, including classification loss (i.e., Equation (3.2)) and regression loss (i.e., Equation (3.3)). The classification loss is the softmax loss over categories $\zeta \in \{vehicle, background\}$. The regression loss is the smooth L_1 loss of the foreground category $\zeta^+ = vehicle$, which regresses to offsets for the center (cx,cy), width (w), and height (h) of the matched anchor box.

$$L_{cls}(c) = -\sum_{i=1}^{N_v} \sum_{\zeta} \lg(c_i^\zeta)$$
(3.2)

$$L_{reg}(p,g) = \sum_{i=1}^{N_v} \sum_{m \in \{cx,cy,w,h\}} \mathbb{I}_{ij}^{\zeta^+} Smooth_{L_1}(p_i^m - g_j^m)$$
(3.3)

where, $\mathbb{I}_{ij}^{\zeta^+} \in \{0,1\}$ is the indicator of whether the i-th anchor box matches the j-th ground-truth box.

The offset and size losses are the smooth L1 loss between the predicted parameters (p) and the ground-truth parameters (g) based on the matched anchor boxes, as shown in Equations (3.4) and (3.5). The vehicle must contain a license plate; otherwise, the losses $L_{\text{off}}(p,g)$ and $L_{\text{size}}(p,g)$ are 0 by setting $g_j^+ = 0$. This way, it can avoid learning falsepositive predictions during training.

$$L_{\text{off}}(p,g) = \sum_{i=1}^{N_v} \sum_{m \in (\text{off}_x, \text{off}_y)} \mathbb{I}_{ij}^{\zeta^+} g_j^+ \text{Smooth}_{L1}(p_i^m - g_j^m) \quad (3.4)$$

$$L_{\text{size}}(p,g) = \sum_{i=1}^{N_v} \sum_{m \in (lp_w, lp_h)} \mathbb{I}_{ij}^{\zeta^+} g_j^+ \text{Smooth}_{L1}(p_i^m - g_j^m) \quad (3.5)$$

where, off_x and off_y are the offsets between the center of the license plate and the vehicle in x-direction and y-direction, lp_w and lp_h are the width and height of the license plate, and $g_j^+ \in \{0,1\}$ is the indicator of whether the j-th vehicle contains a license plate.

Moreover, the probability of the vehicle containing a license plate can be used to reduce false positives of the license plate. A license plate will be detected only when the probability is greater than a certain threshold, and the threshold is empirically set to 0.5. During training, the vehicles with very small-sized or invisible license plates (occlusion, far shooting-distance, etc.) are regarded as without license plates; otherwise, the vehicles are considered as containing a license plate. The containing-LP probability is optimized by the binary cross-entropy loss (i.e., Equation (3.6)).

$$L_{\text{con\_lp}}(p,g) = -\sum_{i=1}^{N_v} [g_i^+ \lg(\sigma(p_i^+)) + (1 - g_i^+) \lg(\sigma(p_i^+))] \quad (3.6)$$

where, σ is a sigmoid function to limit the predicted containing-LP probability $p_i^+ \in [0, 1]$ in case of loss divergence.

After the ALPD stage, there is a large deviation between the predicted license plate and the ground truth. We should make a further refinement to get more precise detection results, i.e., fine-tuning the license plate in the local region around the license plate. Based on the center and size of the license plate, we obtain the local region by merely expanding the license plate region with a preset ratio, enclosing the license plate and little background. The license plate occupies a relatively larger area in the local region than in the input image so that the subsequent refinement network can get more precise detection results. There are many license plate regions obtained from different vehicles simultaneously. All the region features are extracted from the first convolutional layer, ensuring the whole network is constructed in an end-to-end manner. The first convolutional layer preserves the same size as the input image, which retains sufficient spatial information to detect small-sized license plates. Furthermore, all the LP regions are resized and aggregated via differentiable RoI warping for batch operation, ensuring all the license plates are detected simultaneously to reduce the running time.

In the local region, the quadrilateral and horizontal bounding boxes of the license plate are detected simultaneously. The quadrilateral bounding box is obtained by regressing the four corners of the license plate based on the matched anchor box, as illustrated in Fig. 3.3. The matched anchor box is determined by the intersection over union (IoU) with the horizontal ground-truth box. The horizontal bounding box is used for NMS because of the fast computing speed. Compared with the ALPD module, the detection results of the MLPR module are more accurate.

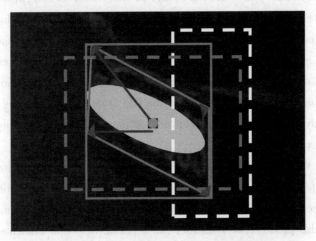

Fig. 3.3 License plate position calculation

The MLPR module has only 6 convolutional layers because the LPD task in the local region is relatively simple. Please refer to our open-source codes for more details. The training objective of the MLPR module is defined as Equation (3.7), including three parts: LP classification loss $L_{cls}(c')$, LP regression loss $L_{reg}(p',g')$, and LP corner loss $L_{corner}(p',g')$.

$$L_2(c',p',g') = \frac{1}{N'_{lp}}[L_{cls}(c') + L_{reg}(p',g') + L_{corner}(p',g')] \quad (3.7)$$

where, N'_{lp} is the number of matched anchor boxes with the horizontal LP ground-truth boxes, c' is the LP presence confidence, p' is the predicted LP parameter, and g' is the LP ground-truth parameter.

The losses of the horizontal bounding box $L_{cls}(c')$ and $L_{reg}(p',g')$ are the same as vehicle detection except for the foreground category being LP, as shown in

Equation(3.2) and Equation(3.3). As shown in Equation(3.8), the corner loss of the quadrilateral bounding box is the smooth L_1 loss of the foreground category $\zeta'^+ =$ license plate, which regresses to offsets between the center of the matched anchor box and the four corners of the license plate.

$$L_{\text{corner}}(p',g') = \sum_{i=1}^{N'_{\text{lp}}} \sum_{m \in \{tl_x,tl_y,tr_x,tr_y,br_x,br_y,bl_x,bl_y\}} \mathbb{I}_{ij}^{\zeta'^+} \text{Smooth}_{L1}(p_i'^m - g_j'^m) \quad (3.8)$$

where, $m \in \{tl_x, tl_y, tr_x, tr_y, br_x, br_y, bl_x, bl_y\}$ are the four corners of the license plate, i.e., top-left, top-right, bottom-right, and bottom-left.

The DLPD module can directly detect the license plate in the input image. In this way, small-sized license plates can not always be detected. However, in some cases, when the license plate fails to be detected in the indirect detection branch due to incorrectly detected vehicles, the DLPD module can reduce the false-negative license plate. The DLPD module is similar to the MLPR module. One significant difference is that the license plate is directly detected in the input image, not in the local region of the license plate. In addition, the backbone network of the DLPD module is the same as SSD with 25 convolutional layers; the backbone network of the MLPR module only consists of 6 convolutional layers.

It is difficult to effectively detect the vehicle and license plate simultaneously due to their subordinate relationships. This issue is caused by feature interaction between the vehicle and license plate in the traditional anchor-based detection method, such as SSD. To solve this problem, we construct two separate detection branches for the DLPD and ALPD modules, respectively, as shown in Fig. 3.2. These two modules share the same backbone network (i.e., VGG-16 and extra layers) but different head networks. Please refer to our open-source codes for more details. In this way, we can eliminate the adverse effects on the license plate caused by the vehicle.

Similar to the MLPR module, the training objective of the DLPD module is defined as Equation(3.9), including LP classification loss $L_{\text{cls}}(c'')$, LP regression loss $L_{\text{reg}}(p'', g'')$, and LP corner loss $L_{\text{corner}}(p'', g'')$.

$$L_3(c'',p'',g'') = \frac{1}{N''_{\text{lp}}}[L_{\text{cls}}(c'') + L_{\text{reg}}(p'',g'') + L_{\text{corner}}(p'',g'')] \quad (3.9)$$

By integrating the indirect and direct detection branches, we develop an end-to-end trainable network for license plate detection, which can effectively detect the small-sized license plate and accurately localize the multidirectional license plate in real applications. Combining Equations (3.1), (3.7), and (3.9), the loss of the whole network is shown in Equation(3.10), where α and β are simply set to 1 to balance these

3.2 Methodology

loss terms.

$$L = L_1(c,p,g) + \alpha L_2(c',p',g') + \beta L_3(c'',p'',g'') \quad (3.10)$$

Fig. 3.4 illustrates the loss changes during training, including L_1 and L_2 of the indirect detection branch as well as L_3 of the direct detection branch. During end-to-end training, the ALPD module can be optimized to detect the vehicle and approximate location of the license plate. Meanwhile, the license plate can be directly detected in the input image by the DLPD module. After training for some iterations, the MLPR module starts to refine the location of the license plate in the local region; then, the entire network will be optimized simultaneously. Specifically, during the first few training iterations, L_1 and L_3 go down, and L_2 remains zero because the untrained ALPD module can not estimate the location of the license plate; then, L_2 goes up dramatically because the ALPD module can approximately localize the license plate, and the MLPR module starts learning to regress the four corners of the license plate in the local region; finally, the total loss L goes down steadily because the indirect and direct detection branches are optimized simultaneously.

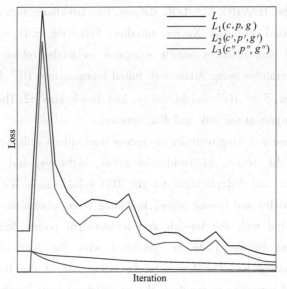

Fig. 3.4 Training loss

Fig. 3.5 illustrates the post-processing operations. We can filter the most useless detection results by thresholding the confidence predicted by the network. After threshold filtering, the post-processing module can merge the detection results from two detection branches via NMS, removing duplicate detections. Instead of the quadrilateral bounding box, the horizontal bounding box of the license plate is used for NMS because of its faster computing speed. The final detection results are mainly from the indirect detection branch because of its ability to detect small-sized license plates. In some cases, the direct detection branch can reduce the false-negative license plate in the indirect detection branch due to incorrectly detected vehicles. In this way, the network can detect the license plate with both high Precision and Recall rates.

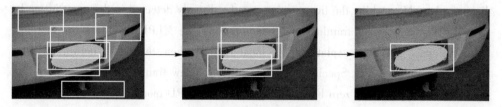

Fig. 3.5　Post-processing operations

3.3　Experiments

The backbone network of the DLPD and ALPD modules follows SSD512, which is initialized with the ILSVRC CLS-LOC dataset. The backbone network of the MLPR module is initialized with the Xavier initializer. Following SSD, we adopt the data augmentation and hard negative mining strategies for model robustness. We train the model for 60 K iterations using Adam with initial learning rate 10^{-4}, $0.9\beta_1$ momentum, $0.99\beta_2$ momentum, 5×10^{-4} weight decay, and batch size 32. The learning rate is decreased by 10 times at the 20k and 40k iterations.

TILT720. We use a driving recorder to capture road videos with a resolution of 720 × 1280, including the scenes of residential areas, highways, and expressways. After keyframe extraction and deduplication, we get 1033 valid images. We carefully annotate all the visible vehicles and license plates, including their subordinate relationships. The vehicle is annotated with the top-left and bottom-right points, forming a horizontal bounding box. The license plate is annotated with the four corners, forming a quadrilateral bounding box. The horizontal bounding box of the license plate is the minimal horizontal bounding rectangle of the quadrilateral bounding box. For simplicity, we name this dataset TILT720 (multidirectional license plate detection dataset 720P). All the images are randomly divided into the training-validation set and test set in the proportion of 9 : 1. TILT1080. Similar to the TILT720, we obtain the TILT1080 with another driving recorder. The TILT1080 contains 4112 images, and all the images have a size of 1080 × 1920. All the images are randomly divided into the training-validation set and test set in the proportion of 9 : 1.

As shown in Table 3.1, we adopt the ALPD module as the benchmark model. The ALPD module is the first step of the indirect detection branch and can approximately estimate the license plate in the input image. The module only achieves very low AP on all the test sets, especially for the IoU threshold 0.75. After only adding the MLPR module, the detection performance worsens because the license plate is refined in the region that cannot completely enclose the license plate. We further add the LREA module, where the license plate region is expanded to 3 times. In this way, the license plate can be refined in the local region that can completely enclose the license plate with a little background. The ALPD, LREA, and MLPR modules assemble the indirect detection branch, improving the AP by 10%-20% with different IoU thresholds compared with the ALPD module.

Table 3.1 Ablation study of different datasets with different IoU thresholds

Method	LREA	MLPR	DLPD	IoU = 0.5		IoU = 0.75	
				TILT720	TILT1080	TILT720	TILT1080
ALPD				76.71%	77.71%	26.27%	35.27%
Indirect		√		40.35%	40.62%	7.48%	10.76%
	√	√		89.19%	87.67%	54.51%	56.92%
Direct			√	86.85%	86.01%	47.52%	53.34%
Two-branch	√	√	√	89.30%	87.79%	56.54%	57.94%

The DLPD module can directly detect the license plate in the input image, which achieves comparable performance with the indirect detection branch with a small IoU threshold; however, with a large IoU threshold, the performance is much lower. The DLPD module cannot accurately localize the license plate in the large input image because of more background noises, making it difficult to detect the small-sized license plate. Combining the indirect and direct detection branches, we get the whole detection network, which achieves higher AP on all the test sets with different IoU thresholds. The network can detect the small-sized license plate via vehicle-plate relation and reduce the false-negative license plate caused by incorrectly detected vehicles. Moreover, the ALPD module, the indirect detection branch, and the whole detection network have almost the same vehicle detection performance as the vanilla SSD. This way, it proves our method can continuously improve the license plate detection performance while maintaining the vehicle detection performance.

We calculate the Precision, Recall, and F1-score based on the predicted quadrilateral bounding box. For the methods that can only detect the horizontal bounding box, we only compare the best SSD with our proposed method. As shown in Table 3.2, our method achieves the best F1-score for all the test sets with different IoU thresholds.

Table 3.2 Comparative experiments of the multidirectional license plate

Method	TILT720(IoU = 0.5/0.75)			TILT1080(IoU = 0.5/0.75)		
	Precision	Recall	F1-Score	Precision	Recall	F1-Score
SSD	98.66/65.10	58.80/38.80	73.68/48.62	93.88/75.92	40.38/30.07	56.47/43.08
Method	88.79/53.27	76.00/45.60	81.90/49.14	83.53/55.08	68.97/45.48	75.55/49.83
SSD+FC	97.47/75.32	61.60/47.60	75.49/58.33	97.57/84.67	42.61/36.98	59.32/51.48
Method	90.61/60.41	88.80/59.20	89.70/59.80	88.17/61.51	87.89/61.32	88.03/61.42
Ours(Direct)	98.69/82.31	60.40/48.40	74.94/60.96	96.96/85.95	44.00/39.00	60.53/53.66
Ours(Indirect)	88.93/60.87	90.00/61.60	89.46/61.23	88.72/61.65	87.78/61.00	88.25/61.32
Ours(Two-branch)	89.68/61.90	90.40/62.40	90.04/62.15	87.85/62.09	89.16/63.02	88.50/62.55

SSD achieves relatively poor performance, because the detection results of SSD have very low IoU with the quadrilateral ground-truth box. Furthermore, like the DLPD module, we upgrade SSD and make it capable of directly detecting the four corners of the license plate in the input image (SSD + FC). SSD + FC can achieve much better performance than the vanilla SSD , especially for the large IoU threshold. However, SSD+ FC suffers low Recall because of the background noises. Our method combines the advantages of method and SSD+FC, and can precisely detect the multidirectional license plate with a higher Recall rate.

Some qualitative detection results are illustrated in Fig. 3.6. The license plate can be detected via vehicle-plate relation in the indirect detection branch, especially for the small-sized license plate. However, when many vehicles are close to each other, some vehicles may be detected with a large deviation, as shown in the first two images. In addition, in some cases, the vehicle fails to be detected due to boundary truncation, as shown in the third image. In these cases, the license plate cannot be detected in the indirect detection branch. Meanwhile, the license plate can be directly detected in the input image in the direct detection branch. However, due to the disturbing background noises, the direct detection branch can only detect relatively large and horizontal license plate.

By combing these two detection branches with post-processing operations, such as NMS, we get the final detection results. As can be seen, these two detection branches are complementary to each other. The indirect detection branch can detect most of the license plates; in some cases, the direct detection branch can reduce the false-negative license plate in the indirect detection branch due to undetected vehicles or vehicles with large deviations.

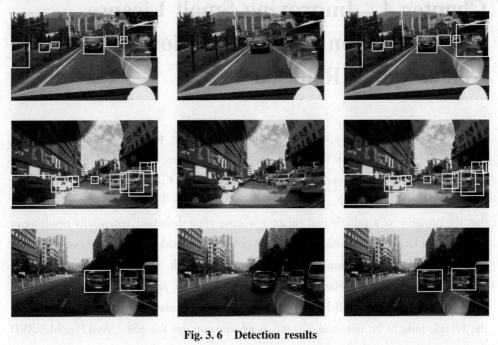

Fig. 3.6 Detection results

Please see colorful images

Chapter 4　Improving Small License Plate Detection with Bidirectional Vehicle-plate Relation

License plate detection is a critical component of license plate recognition systems. A challenge in this domain is detecting small license plates captured at a considerable distance. Previous researchers have proved that pre-detecting the vehicle can enhance small license plate detection. However, this approach only utilizes the one-way relation that the presence of a vehicle can enhance license plate detection, potentially resulting in error accumulation if the vehicle fails to be detected. To address this issue, we propose a unified network that can simultaneously detect the vehicle and the license plate while establishing bidirectional relationships between them. The proposed network can utilize the vehicle to enhance small license plate detection and reduce error accumulation when the vehicle fails to be detected. Extensive experiments on the SSIG, AOLP, and CRPD datasets prove our method achieves state-of-the-art detection performance, achieving an average detection $AP_{0.5}$ of 99.5% on these three datasets, especially for small license plates. When incorporating a license plate recognizer that relies on character detection, we can achieve an average recognition accuracy of 95.9%, surpassing all comparative methods.

4.1　Problem formulation

Automatic license plate recognition (ALPR) has recently gained significant popularity in various applications, such as traffic enforcement, theft detection, and automatic toll collection. The ALPR system typically consists of three stages: license plate detection, character detection, and character recognition. Among these stages, license plate detection plays a pivotal role in determining the overall accuracy of the ALPR system. Specifically, detecting small license plates presents a significant challenge due to their size.

As shown in Fig. 4.1(a), many ALPR methods have been proposed to directly detect the license plate from the input image. However, detecting the license plate directly can lead to missed detections, primarily due to its small size. To address this issue, Kim et al. propose a two-step approach as depicted in Fig. 4.1(b), where the vehicle is first pre-detected, followed by license plate detection within the vehicle region. These methods reduce the search region and mitigate background noises, enhancing license plate detection. Nevertheless, these methods may encounter error accumulation if the vehicle fails to be detected, resulting in subsequent failures in license plate detection. To minimize error accumulation, Chen et al. propose a fusion approach illustrated in Fig. 4.1(c), which combines direct license plate detection(Fig. 4.1(a)) and vehicle pre-detection (Fig. 4.1(b)), merging both detection branches to obtain the final results. However, this approach is time-consuming due to the involvement in multiple detection branches and the subsequent merge operation.

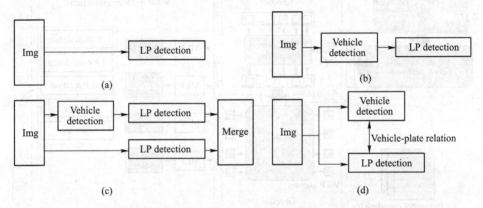

Fig. 4.1 Four methods for license plate detection

To address the challenges mentioned earlier, as depicted in Fig. 4.1(d), we propose simultaneous detection of both the vehicle and the license plate, leveraging their bidirectional relationship to enhance small license plate detection. This approach facilitates mutual reinforcement between vehicles and license plates due to their interdependency. In comparison to direct detection(Fig. 4.1(a)), our method utilizes the presence of the vehicle to improve license plate detection. Unlike the vehicle pre-detection approach(Fig. 4.1(b)), our method mitigates error accumulation arising from the one-way relationship between the vehicle and the license plate. Additionally, compared to the fusion approach(Fig. 4.1(c)), our method enhances inference speed through simultaneous detection and bidirectional relation mining. Extensive experiments on the SSIG, AOLP, and CRPD datasets validate the effectiveness of our method, achieving an average detection $AP_{0.5}$ of 99.5%, particularly for small license plates. When combined with a YOLO-based character recognizer, our method outperforms other state-of-the-art techniques, achieving an average recognition accuracy of

95.9%. Notably, annotations for both vehicles and license plates are available for the SSIG and AOLP datasets within the community. However, for the CRPD dataset, only license plate annotations are provided.

4.2 Methodology

As depicted in Fig. 4.2, our proposed network can simultaneously detect vehicles and license plates and generate their bidirectional relationships. When a license plate subordinates to a vehicle, their relation confidence is higher, and vice versa. This way, it can mutually enhance the detection of vehicles and license plates.

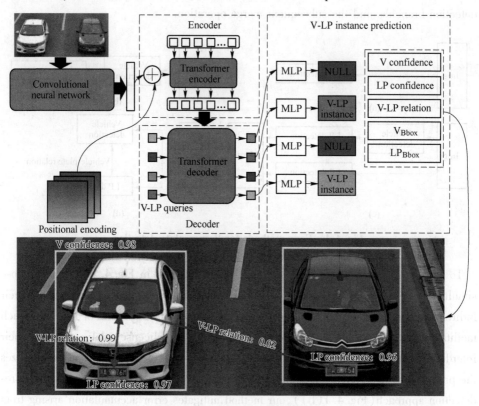

Fig. 4.2　Overall architecture

4.2.1　Network architecture

The proposed network can be mainly divided into three parts: (Ⅰ) A CNN backbone to extract visual features from the input image; (Ⅱ) A transformer encoder-decoder to process visual features and generate global features; (Ⅲ) A multi-layer perceptron layer(MLP) to generate predictions based on global features.

Backbone: We utilize ResNet-50 to extract visual features from the input image into feature maps. The size of the input image and features maps is $[H_0, W_0, 3]$ and $[H, W, C]$ respectively, s. t. , $H = H_0/32$ and $W = W_0/32$ Subsequently, a 1 × 1 convolutional layer is utilized to reduce the channel dimension from $C = 2048$ to $d = 256$. Since the subsequent encoder requires a sequence as input, we convert the reduced features into a sequence of length $H \times W$ where each step corresponds to a vector of size d. As a result, we obtain a flattened feature map with the dimension of $[H \times W, d]$.

Encoder: The encoder follows the vanilla transformer, incorporating six identical units. Each unit comprises an eight-head self-attention network and a two-layer feed-forward network (FFN) with the dimension of $d_{ff} = 2048$. The output dimension is set to $d_{model} = 512$. The Query, Key, and Value are all obtained by the sum of positional encodings and visual features from the CNN backbone to generate global features.

Decoder: The decoder also follows the vanilla transformer, incorporating six identical units. Each unit comprises an eight-head cross-attention network, an eight-head self-attention network, and a two-layer feed-forward network. Similar to the encoder, the FFN dimension is $d_{ff} = 2048$ and the output dimension is $d_{model} = 512$. The decoder takes three inputs, i. e. , positional encodings, V-LP queries, and global features from the encoder, to generate $N = 100$ embeddings for predictions. In the cross-attention network, the Value is obtained directly from global features. The Key is the sum of global features and positional encodings, and the Query is the sum of positional encodings and V-LP queries.

Vehicle-plate Instance Prediction: The output embeddings generated by the decoder are converted into vehicle-plate instances using MLPs. We define the vehicle-plate instance as a five-tuple consisting of vehicle confidence, vehicle-plate relation confidence, plate confidence, vehicle box, and plate box. Specifically, two three-layer MLPs are employed to predict the bounding box of the vehicle and the license plate. Additionally, three single-layer MLPs are utilized to estimate the confidence of the vehicle, the plate, and the vehicle-plate relation.

4.2.2 Training objective

We treat the prediction of vehicle-plate instances as a problem of set prediction, involving a bipartite matching between the predicted instances and the ground truth. When presented with an input image, our model generates $N = 100$ predicted instances, where N represents the number of V-LP queries. The prediction set is represented as $P = P^i, i = 1, 2, \cdots, N$. The ground-truth set is represented as $G = g^i, i = 1,$

$2,\cdots,M,\phi,\cdots,\phi$, where ϕ denotes a null value for one-to-one matching between P and G, and M denotes the total number of ground-truth instances, s. t., $M \leqslant N$. The number of ϕ plus M equals N.

As demonstrated in Equation (4.1), we use the Hungarian algorithm to find the best bipartite matching $\hat{\sigma}$ by minimizing the overall matching cost ζ_{cost}, which is composed of the matching cost of all N matching pairs.

$$\hat{\sigma} = \arg\min \zeta_{\text{cost}}, \sigma \in \mathcal{O}_N$$
$$\zeta_{\text{cost}} = \sum_i^N \zeta_{\text{match}}(g^i, p^{\sigma(i)}) \quad (4.1)$$

where, $\mathcal{O}_N$ represents the one-to-one matching solution space, and σ represents an injective function from the ground-truth set G to the prediction set P. $\zeta_{\text{match}}(g^i, p^{\sigma(i)})$ represents the matching cost between the i-th ground truth and $\sigma(i)$-th prediction, where $\sigma(i)$ represents the matching index of the prediction.

As demonstrated in Equation (4.2), the matching cost of each pair contains the classification loss ζ_{cls}^j and bounding box regression loss ζ_{box}^k.

$$\zeta_{\text{match}}(g^i, p^{\sigma(i)}) = \beta_1 \sum_{j \in v,p,r} \alpha_j \zeta_{\text{cls}}^j + \beta_2 \sum_{k \in v,p} \zeta_{\text{box}}^k \quad (4.2)$$

where, v, p, r represents the vehicle, license plate, and vehicle-plate relation, respectively. ζ_{cls}^j is calculated by the softmax cross-entropy loss. ζ_{box}^k is calculated by the weighted sum of L_1 loss and GIoU loss. In this work, we emphasize classification by setting β_1 to 2 and β_2 to 1. Among classification, we emphasize vehicle-plate relation by setting α_r to 2, α_v to 1, and α_p to 1.

The ground truth during training is illustrated in Fig. 4.3.

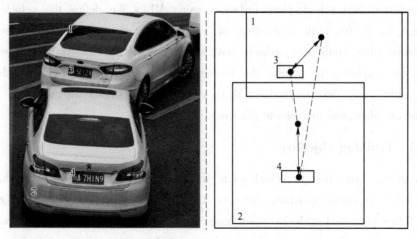

Fig. 4.3 Ground truth during training

The red and green boxes denote the ground-truth boxes of vehicles and license plates, respectively. The solid purple line represents the ground-truth V-LP relation, i. e., the positive relation sample used during training. The dotted purple line denotes no relation between the vehicle and the license plate, i. e., the negative relation sample, which is not used during training.

4.3 Experiments

4.3.1 Datasets

We utilize three publicly available datasets: SSIG, AOLP, and CRPD. SSIG and AOLP provide the annotations for the vehicle and the license plate, but CRPD only provides the annotations for the license plate.

SSIG comprises 2000 Brazilian license plates obtained from 101 vehicles. Following the official settings, we use 40% images for training, 20% for validation, and 40% for testing.

AOLP consists of three distinct subsets, each captured using different shooting methods. The AC subset focuses on static vehicles, while the LE subset captures vehicles violating traffic rules via roadside cameras. The RP subset captures images from various viewpoints and distances using cameras mounted on patrol vehicles. In total, the dataset includes 2049 images containing Chinese Taiwanese license plates. When testing on one subset, the other two subsets are used for training and validation.

CRPD has 33757 Chinese license plates captured by overpasses, which cover various vehicle models, such as cars, trucks, and buses. We follow the official split, i. e., 25000 images for training, 6250 for validation, and 2300 for testing.

4.3.2 Valuation protocols

We use Average Precision (AP) to evaluate license plate detection. Specifically, we utilize the computation method introduced in COCO that calculates AP with different IoU(Intersection over Union) thresholds, i. e., ranging from 0.5 to 0.95 with an interval of 0.05. $AP_{0.5}$ refers to the average precision calculated at the IoU threshold of 0.5. We utilize Accuracy as the evaluation metric for license plate recognition, where all characters must be recognized accurately. We use Frame Per Second(FPS) to calculate the inference speed.

In addition, to verify the effectiveness of small license plate detection, we categorize license plates into three groups based on their height. License plates with a height of 25 pixels or less are categorized as small(S), those exceeding 25 pixels but not exceeding

50 pixels are categorized as medium (M), and license plates taller than 50 pixels are categorized as large (L).

4.3.3 Ablation study

As presented in Table 4.1, Table 4.2, and Table 4.3, we investigate the impact of implicit and explicit relationships between vehicles and license plates on the SSIG, AOLP, and CRPD datasets, respectively. We conduct three ablation experiments: (1) Direct license plate detection using the vanilla DETR model; (2) Simultaneous vehicle and license plate detection using the vanilla DETR model, which implicitly captures the relation between vehicles and license plates; (3) Our proposed method, except for simultaneous vehicle and license plate detection, explicitly incorporating vehicle-plate relationships. After performing license plate detection, we employ the same YOLO-based character recognizer for license plate recognition. Implicit vehicle-plate relationships have minimal impact on license plate detection and recognition. However, when incorporating explicit vehicle-plate relationships, our method substantially improves license plate detection and recognition. Additionally, our method enhances vehicle detection due to the bidirectional relationships between vehicles and license plates.

Table 4.1 Ablation study on SSIG

Method	LP	V	Relation	Detection(V)		Detection(LP)		Recognition
				AP	$AP_{0.5}$	AP	$AP_{0.5}$	Accuracy
DETR	√			—	—	45.6%	96.3%	95.4%
	√	√		78.0%	99.2%	50.1%	97.5%	95.6%
Ours	√	√	√	81.4%	100.0%	60.6%	100.0%	96.4%

Table 4.2 Ablation study on AOLP

Method	LP	V	R	Detection(V)			Detection(LP)			Recognition		
				AP			AP			Accuracy		
				AC	LE	RP	AC	LE	RP	AC	LE	RP
DETR	√			—	—	—	53.2%	52.2%	43.6%	96.1%	94.3%	95.6%
	√	√		89.2%	87.8%	83.6%	52.2%	54.6%	40.6%	96.2%	95.0%	94.5%
Ours	√	√	√	93.9%	90.8%	91.5%	65.2%	60.8%	58.2%	98.1%	98.0%	97.6%

Table 4.3 Ablation study on CRPD

Method	LP	V	Relation	Detection(V)		Detection(LP)		Recognition
				AP	$AP_{0.5}$	AP	$AP_{0.5}$	Accuracy
DETR	√			—	—	53.0%	96.3%	86.0%
	√	√		83.9%	98.5%	54.2%	96.2%	87.5%
Ours	√	√	√	87.2%	98.6%	62.8%	98.6%	89.3%

4.3.4 Comparative experiments

As presented in Table 4.4, Table 4.5, and Table 4.6, we conduct comparative experiments on the SSIG, AOLP, and CRPD datasets, respectively. In all of these datasets, we compare three approaches: direct detection (Fig. 4.1(a)), vehicle pre-detection(Fig. 4.1(b)), and two branches combining direct detection and vehicle pre-detection(Fig. 4.1(c)). To ensure a fair comparison, all of these comparative methods utilize the same backbone and transformer as our proposed method. After performing license plate detection, both the comparative methods and our proposed method employ the same YOLO-based character recognizer for license plate recognition.

Our proposed method demonstrates superior detection and recognition performance on the SSIG and CRPD datasets while achieving the best performance on most subsets within the AOLP dataset. Concretely, our proposed method achieves an average $AP_{0.5}$ of 99.5% and an average recognition accuracy of 95.9% on the SSIG and CRPD datasets and three subsets of AOLP. However, for the LE subset of AOLP, our proposed method can not effectively handle some low-light images. In future work, we aim to enhance license plate detection under low-light conditions.

Moreover, the direct detection method offers the fastest inference speed but suffers from the lowest detection and recognition performance due to its limited ability to detect small license plates. On the other hand, the vehicle pre-detection method improves license plate detection at the cost of slower inference speed. By combining direct detection and vehicle pre-detection, the two branches method further enhances license plate detection and recognition, albeit with the slowest inference speed. In contrast, our proposed method achieves the best detection and recognition performance while maintaining a comparable inference speed to the direct detection method.

Table 4.4 Comparative experiments on SSIG

Method	Detection		Recognition	
	AP	$AP_{0.5}$	FPS	Accuracy
RARE	—	—	—	93.7%
Rosetta	—	—	—	94.3%
Direct detection	45.6%	96.3%	**13.0**	95.4%
Vehicle pre-detection	52.6%	97.5%	7.7	95.6%
STAR-Net	—	—	—	96.1%
Two branches	—	—	—	96.1%
Ours	**60.6%**	**100.0%**	12.2	**96.4%**

Table 4.5 Comparative experiments on AOLP

Method	Detection						Recognition		
	AC		LE		RP		AC	LE	RP
	AP	$AP_{0.5}$	AP	$AP_{0.5}$	AP	$AP_{0.5}$	Accuracy		
RCLP	—	98.5%	—	97.8%	—	95.3%	94.8%	94.2%	88.4%
DLS	—	92.6%	—	93.5%	—	92.9%	96.2%	95.4%	95.1%
DELP	—	99.3%	—	**99.2%**	—	99.0%	97.8%	97.4%	96.3%
Direct detection	53.2%	98.2%	52.2%	96.1%	43.6%	97.8%	96.1%	94.3%	95.3%
Vehicle pre-detection	47.8%	98.1%	53.8%	96.3%	44.4%	96.9%	96.2%	95.0%	94.5%
Two branches	58.4%	96.4%	57.8%	93.5%	48.8%	98.2%	94.7%	92.2%	96.2%
Ours	**65.2%**	**100.0%**	**60.8%**	99.0%	**58.2%**	**100.0%**	**98.1%**	**98.0%**	**97.6%**

Table 4.6 Comparative experiments on CRPD

Method	Detection		Recognition	
	AP	$AP_{0.5}$	FPS	Accuracy
SYOLOv4+CRNN	—	—	—	71.0%
RCNN+CRNN	—	—	—	73.7%
UCLP	—	—	—	84.1%
Direct detection	53.0%	96.3%	**12.8**	86.0%
Vehicle pre-detection	57.4%	97.8%	7.4	86.2%
Two branches	58.8%	98.1%	4.8	87.5%
Ours	**62.9%**	**98.3%**	12.5	**89.3%**

Fig. 4.4 demonstrates that our proposed method can accurately detect vehicles and license plates, and the YOLO-based character recognizer can accurately recognize the

detected license plates based on character detection.

Fig. 4.4 Visualization examples of license plate detection and recognition

Table 4.7 presents comparative experiments involving multi-scale license plates on the CRPD dataset.

Table 4.7 Comparative experiments on multi-scale license plates of the CRPD dataset

Method	Detection(LP)						Recognition		
	S		M		L		S	M	L
	AP	$AP_{0.5}$	AP	$AP_{0.5}$	AP	$AP_{0.5}$	Accuracy		
Direction detection	43.3%	92.4%	56.7%	96.5%	62.6%	96.9%	82.2%	86.1%	86.8%
Simultaneous detection	45.0%	92.0%	56.6%	96.8%	62.1%	96.9%	82.0%	86.2%	86.7%
Vehicle pre-detection	48.7%	93.5%	59.0%	97.3%	62.4%	98.0%	83.5%	87.4%	87.6%
Two branches	50.5%	93.6%	60.4%	98.1%	64.7%	98.5%	84.0%	88.4%	88.5%
Ours	55.0%	95.6%	62.5%	98.4%	67.3%	99.2%	85.1%	89.2%	89.9%

Notably, we do not conduct multi-scale experiments on the SSIG and AOLP datasets because the size of license plates in these datasets is relatively consistent. In all of these sizes, we compare three approaches: direct detection (Fig. 4.1(a)), vehicle pre-detection(Fig. 4.1(b)), and two branches combining direct detection and vehicle pre-detection(Fig. 4.1(c)). Moreover, the simultaneous detection method denotes detecting vehicles and license plates simultaneously using the vanilla DETR model. To ensure a fair comparison, all of these comparative methods utilize the same backbone and transformer as our proposed method. After performing license plate detection, both the comparative methods and our proposed method employ the same YOLO-based character recognizer for license plate recognition. Our proposed method demonstrates superior performance in both license plate detection and recognition across all sizes, especially for small license plate detection. Concretely, it achieves a 4.5% AP improvement in the detection performance of small license plates compared to the two branches method, with a 2.1% AP improvement for medium license plates and a 2.6% AP improvement for large license plates.

As depicted in Fig. 4.5, our method can effectively detect small license plates at a considerable distance. Our method can achieve comparative inference speed with the direct detection method, surpassing other comparative methods. Moreover, our method can detect vehicles truncated by image edges due to the bidirectional relationships between vehicles and license plates.

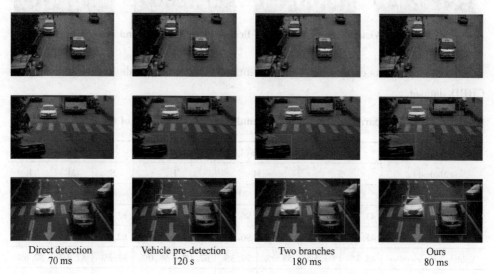

| Direct detection | Vehicle pre-detection | Two branches | Ours |
| 70 ms | 120 s | 180 ms | 80 ms |

Fig. 4.5 Visualization examples

Please see colorful images

Chapter 5　Vertex Adjustment Loss for Multidirectional License Plate Detection and Recognition

Due to various shooting angles, the license plates are captured multidirectional in the wild. The license plates will only undergo rigid deformation in the image because of their rigid body property. Based on this fact, many researchers propose to detect the multidirectional license plate by regressing its vertices. However, these methods regard the regression of each vertex as an independent task and ignore the relationship between vertices. Since the regression distance of each vertex differs, it can cause different deviations of the regressed vertices and affect the subsequent recognition. To solve this problem, we propose to utilize the relationship between vertices to improve vertex prediction for multidirectional license plate detection. Specifically, besides vertex regression, we propose to adjust the regressed vertices by aligning the minimal rectangle formed by them to the ground-truth box. This way, the deviated vertices can be adjusted closer to the ground-truth vertices for accurate detection. The vertex adjustment is only performed in the training phase by adding a costless loss, increasing no network parameters. Extensive experiments verify our proposed method can improve license plate detection and recognition on the CCPD, CLPD, and RodoSol-ALPR datasets, especially for the highly tilted license plates. Moreover, experiments on the scene text dataset ICDAR2015 prove its effectiveness for skewed text detection.

5.1　Problem formulation

Accurate license plate detection is essential to the subsequent recognition. However, the license plates captured in the wild are multidirectional due to various shooting angles, making it challenging to detect in real scenarios. Based on the prestigious object detectors, such as Faster-RCNN, SSD, and YOLOv2, many specific methods are proposed for license plate detection. However, these methods are designed to detect horizontal rectangles and are unsuitable for multidirectional license plates. Hence, considering the rigid-body property of the license plate, recent methods can be used to detect the multidirectional. license plate by regressing its vertices. However, these methods predict each vertex independently and do not consider the relationship between vertices. As shown in Fig. 5.1(a), the regression distances between the base point and four vertices are different, causing deviations of the predicted vertices. Inaccurate vertex prediction will lead to errors in the subsequent shape rectification and recognition. *represents missing characters.

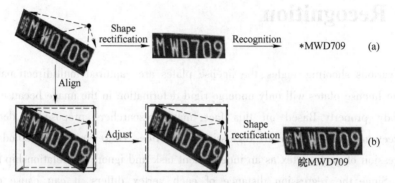

Fig. 5.1 License plate vertex prediction and correction

To solve this problem, we propose to utilize the relationship between the predicted vertices to improve vertex prediction. As shown in Fig. 5.1(b), besides vertex regression, we propose to align the minimal rectangle formed by the regressed vertices to overlap with the ground-truth box. This way, the deviated vertices can be adjusted closer to the ground-truth vertices for accurate detection, thus improving the subsequent license plate recognition. Specifically, we use IoU (Intersection over Union) to measure the overlap degree of two rectangles and adopt IoU loss to align them. By adding the vertex adjustment loss during training, we can improve multidirectional license plate detection without in-creasing network parameters and inference time. Experiments prove the proposed vertex adjustment loss can be plug-and-play to enhance multidirectional license plate detection and recognition. Compared with the baseline models, our proposed method can improve an average of 1.46% detection F1-score and 2.05% recognition accuracy on the license plate datasets CCPDv2, CLPD, and RodoSol-ALPR with a faster inference speed. We can achieve near 96% recognition accuracy on CCPDv1, an older version of CCPD, surpassing many state-of-the-art methods. Moreover, we can improve skewed text detection by a 1.78% F1-score on the scene text dataset ICDAR2015.

5.2 Methodology

The overall architecture is illustrated in Fig. 5.2, including vertex prediction, shape rectification, and recognition. The vertex prediction network comprises a backbone and detection head, where the backbone follows the anchor-based one-stage SSD for multi-scale object detection using multi-level features. We mainly modify the detection head to improve multidirectional license plate detection. First, we propose to utilize the relationship between the predicted vertices to improve vertex prediction, i. e., aligning the minimal rectangle formed by them to overlap with the ground-truth box using CIoU loss. The vertex adjustment is only performed during the training phase without increasing network pa-rameters and inference time. Second, previous methods, can simultaneously detect the bounding box and vertices of the license plate. We propose to remove bounding box detection, keeping only vertex regression for more accurate and faster detection. The minimal rectangle formed by the adjusted vertices can be regarded as the horizontal bounding box. Moreover, the shape rectification is implemented by the perspective transformation in OpenCV. The recognition is implemented by LPRNet, a lightweight license plate recognition framework without preliminary character segmentation and complex RNNs1.

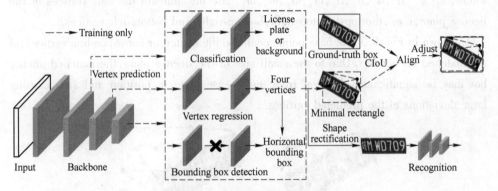

Fig. 5.2 Overall architecture

The baseline models, can simultaneously detect the bounding box and vertices of the license plate. The training loss comprises classification loss L_{cls}, bounding box detection loss L_{bb}, and vertex regression loss L_{vr}.

$$L_{base} = \frac{1}{N}(L_{cls} + \alpha L_{bb} + \beta L_{vr}) \tag{5.1}$$

Then, the tilted license plate can be transformed horizontally based on the adjusted vertices and recognized to obtain the license plate characters. Where N is the number of matched anchor boxes, and α and β are the loss balance terms.

The classification loss L_{cls} is the Softmax loss of classes $c \in \{$license plate, background$\}$, where p is the class confidence.

$$L_{\text{cls}} = -\sum_{i=1}^{N}\sum_{c} \lg(p_i^c), \quad p_i^c = \frac{\exp(\bar{p}_i^c)}{\sum_{c}\exp(\bar{p}_i^c)} \quad (5.2)$$

The bounding box detection loss L_{bb} is the Smooth $L1$ loss of the foreground category $c^+ = $ license plate, which regresses to offsets for the center (cx, cy), width (w), and height (h) of the matched anchor box.

$$L_{\text{bb}} = \sum_{i=1}^{N}\sum_{m \in \{cx,cy,w,h\}} \mathbb{I}_{ij}^{c^+} \text{Smooth}_{L1}(l_i^m - g_j^m) \quad (5.3)$$

where, $\mathbb{I}_{ij}^{c^+} \in \{0,1\}$ is the indicator of whether the i-th anchor box matches the j-th ground-truth box. l is the predicted offsets, and g is the ground-truth offsets.

The vertex regression loss L_{vr} is the Smooth $L1$ loss to predict the offsets between the base points and vertices of the license plate.

$$L_{\text{vr}} = \sum_{i=1}^{N}\sum_{m} \mathbb{I}_{ij}^{c^+} \text{Smooth}_{L1}(l_i^m - g_j^m) \quad (5.4)$$

where, $m \in \{tl_x, tl_y, tr_x, tr_y, br_x, br_y, bl_x, bl_y\}$ are the four are the four vertices of the license plate, i.e., the top-left, top-right, bottom-right, and bottom-left vertices.

As shown in Fig. 5.3, the base points could be the center, or corresponding vertices, of the matched anchor box. Due to the small size of the license plate, the matched anchor box may be significantly different from the license plate in position and size, causing large deviations of the predicted vertices.

(a) (b)

Fig. 5.3 Vertex regression based on the center

Based on the baseline models, we propose to remove bounding box detection L_{bb} and add a vertex adjustment loss $L_{\text{va}}(B, B_{\text{gt}})$.

$$L = \frac{1}{N}[L_{\text{cls}} + L_{\text{vr}} + \gamma L_{\text{va}}(B, B_{\text{gt}})] \quad (5.5)$$

where, B_{gt} is the ground-truth box, and γ is the loss balance term, which is set to 16 by default. $B = (x_{\min}, y_{\min}, x_{\max}, y_{\max})$ is the minimal rectangle formed by the predicted vertices.

$$\begin{aligned} x_{\min} = \min(tl_x, bl_x), \quad y_{\min} = \min(tl_y, tr_y) \\ x_{\max} = \max(tr_x, br_x), \quad y_{\max} = \max(bl_y, br_y) \end{aligned} \quad (5.6)$$

$L_{va}(B,B_{gt})$ is implemented by CIoU loss, which considers the overlap area, central point distance, and aspect ratio to align two rectangles.

$$L_{va}(B,B_{gt}) = 1 - \text{IoU} + \frac{E^2(c,c_{gt})}{d^2} + \delta V$$

$$\text{IoU} = \frac{B \cap B_{gt}}{B \cup B_{gt}}, \quad \delta = \frac{V}{(1-\text{IoU})+V} \tag{5.7}$$

$$V = \frac{4}{\pi^2}\left(a\tan\frac{w}{h} - a\tan\frac{w_{gt}}{h_{gt}}\right)^2$$

where, IoU measures the overlap degree of two boxes, E denotes the Euclidean distance between the center of two boxes, and measures the consistency of the aspect ratio, d denotes the diagonal length of the smallest enclosing box covering B and B_{gt}.

5.3 Experiments

The detection models are trained for 60k iterations using the Adam optimizer with initial learning rate 10^{-4}, β_1 momentum 0.9, β_2 momentum 0.99, weight decay 5×10^{-4}, and batch size 32. The learning rate is decreased by 10× at 20k and 40k iterations.

5.3.1 Datasets

Chinese City Parking Dataset (CCPD) is collected in the roadside parking scenes, including two versions: C-CPDv1 and CCPDv2. The image size is 720 1160. CCPDv1, containing more than 250k images, is divided into several subsets, i.e., Base (basic subset), DB (various illumination), FN (different shooting distance), Rotate (in-plane rotation), Tilt (out-of-plane rotation), Weather (different weather), and Challenge (challenging conditions). Compared with CCPDv1, CCPDv2 contains more than 300k images with an extra subset Blur (blurry images) without the subset Weather. The subset Base is used for training and validation, and the other subsets are used for testing. We choose CCPDv2 as the benchmark dataset because it is more challenging and representative than CCPDv1.

China License Plate Dataset (CLPD) is collected from various real-world scenes, such as the Internet, mobile phones, and driving recorders, which contains 1200 images from all 31 provinces in mainland China. The images in CLPD are captured with various shooting angles, image resolutions, and backgrounds. We randomly split the CLPD dataset following the classic 70% : 30% strategy, i.e., 840 images for training and validation, 360 images for testing.

RodoSol-ALPR is collected at toll booths installed on a Brazilian highway, containing

20k images. RodoSol-ALPR has images of two different layouts: Brazilian and Mercosur 2, and half of the vehicles are motorcycles. The image size is 1280×720. We follow the official split, i. e. , 8000, 4000, and 8000 images for training, validation, and testing, respectively. The license plates in CCPD, CLPD, and RodoSol-ALPR are carefully annotated, including four vertices and characters.

ICDAR2015 is obtained with wearable cameras, containing 1500 real scene images. The image size is 1280×720, and all the text regions are annotated with four vertices for skewed text detection. We follow the official split, i. e. , 1000 images for training and validation and 500 for testing.

5.3.2 Evaluation metrics

For detection, we adopt AP (average precision) and F1-score to evaluate the horizontal and quadrilateral bounding box of the license plate, respectively. $AP_{0.5:0.95}$ (abbr. AP) and $F1_{0.5:0.95}$ (abbr. F1) denote the average AP and F1-score of different IoU thresholds from 0.5 to 0.95, where the IoU is calculated as Fig. 5.4. Moreover, we use frames per second (FPS) to evaluate the inference speed. For recognition, we use Accuracy, ED-1, Chinese, and w/o Chinses to to evaluate the accuracy of the entire license plate, edit distance $\leqslant 1$, the Chi-es character, and the characters expect for the Chinese Character, respectively.

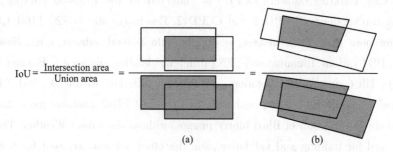

Fig. 5.4 IoU of two horizontal boxes for AP(a) and IoU of two quadrilateral boxes for F1-score(b)

5.3.3 Ablation study

As shown in Table 5.1, we conduct ablation experiments on CCPDv2 with different detection heads, summarized as follows.

The number in the parentheses denotes the weight of IoU loss, i. e. , γ in Equation(5.5).

5.3 Experiments

Higher F1-score means more accurate detection of the quadrilateral bounding box of the multidirectional license plate. Our proposed FV+CIoU achieves the best F1-score, thus achieving the best recognition accuracy in Table 5.3, Table 5.5. Meanwhile, FV+CIoU achieves comparable AP and inference speed with BB and BB(CIoU). However, BB and BB(CIoU) perform a very low F1-score because the predicted horizontal rectangle is far different from the quadrilateral ground-truth box, causing low recognition ac-curacy. The network without BB can improve the F1-score because the network can focus on vertex regression, such as FV vs. BB+FV. Moreover, we experiment with the weight of CIoU loss, i.e., the balance term γ in Equation(5.5). FV+CIoU can improve the F1-score compared with FV and achieve the best F1-score with the weight of 16. However, it will reduce the F1-score with a large weight due to over-fitting. We also prove CIoU loss is better than IoU loss, GIoU loss, and DIoU loss for vertex adjustment because it simultaneously considers the overlap area, central point distance, and aspect ratio.

Table 5.1 Ablation study on CCPDv2

Method	Validation		Test		FPS
	F1	AP	F1	AP	
BB	62.50%	95.77%	37.87%	80.63%	55.7
BB(CIoU)	62.45%	96.57%	38.28%	81.05%	55.7
BB+FV	94.09%	94.80%	77.79%	78.16%	51.2
FV	94.13%	95.02%	77.87%	78.99%	53.6
BB+FV+CIoU	94.06%	94.53%	76.72%	77.51%	51.2
BB(CIoU)+(FV+CIoU)	95.28%	96.58%	78.12%	80.33%	51.2
FV+CIoU(1)	94.37%	95.14%	77.77%	79.06%	53.6
FV+CIoU(3)	94.55%	95.23%	78.13%	78.80%	53.6
FV+CIoU(5)	94.62%	95.46%	78.13%	79.46%	53.6
FV+CIoU(10)	95.08%	95.78%	78.25%	80.03%	53.6
FV+CIoU(16)	95.34%	95.95%	78.69%	80.07%	53.6
FV+CIoU(30)	95.24%	96.93%	78.28%	80.03%	53.6
FV+CIoU(50)	95.17%	96.35%	78.13%	80.86%	53.6
FV+CIoU(100)	93.53%	96.56%	74.42%	79.95%	53.6

Continued Table 5.1

Method	Validation		Test		FPS
	F1	AP	F1	AP	
FV+IoU(16)	95.09%	96.37%	78.17%	79.69%	53.6
FV+GIoU(16)	95.05%	95.95%	78.60%	79.69%	53.6
FV+DIoU(16)	95.11%	96.49%	78.45%	79.82%	53.6

BB: only detecting the horizontal bounding box of the license plate, i.e., the vanilla SSD.

BB(CIoU): replacing the bounding box detection loss(Equation(5.3)) of BB from Smooth L_1 loss to CIoU loss.

BB+FV: simultaneously detecting the horizontal bounding box and four vertices of the license plate.

FV: our proposed method that only detects four vertices of the license plate.

BB+FV+CIoU: our proposed method that uses CIoU loss to align the two boxes obtained by BB and FV.

BB(CIoU)+(FV+CIoU): the combination of BB(CIoU) and our proposed FV+CIoU.

FV+CIoU: our proposed method as Fig. 5.2.

5.3.4 Comparative experiments

As shown in Table 5.2, we conduct comparative experiments on CCPDv2 using SSD-style detectors, of different backbones, input sizes, and regression strategies.

Table 5.2 Comparative experiments on CCPDv2

Backbone	Input size	Vertex regression	Detection head	$F1_{0.5:0.95}$	$F1_{0.9}$
VGGNet	300	None	BB	37.40%	8.24%
			BB(CIoU)	37.64%	9.37%
		Center	BB+FV	74.36%	40.03%
			FV+CIoU	75.55%	43.80%
		Vertex	BB+FV	74.43%	39.99%
			FV+CIoU	75.48%	43.52%
	512	None	BB	37.87%	9.21%
			BB(CIoU)	38.28%	9.77%
		Center	BB+FV	77.79%	47.37%
			FV+CIoU	78.69%	51.12%
		Vertex	BB+FV	77.23%	47.07%
			FV+CIoU	78.23%	50.36%

Continued Table 5.2

Backbone	Input size	Vertex regression	Detection head	$F1_{0.5:0.95}$	$F1_{0.9}$
MobileNet	300	None	BB	37.73%	7.33%
			BB(CIoU)	38.29%	5.84%
		Center	BB+FV	68.81%	28.62%
			FV+CIoU	70.28%	30.64%
		Vertex	BB+FV	66.68%	25.33%
			FV+CIoU	67.88%	28.86%
	512	None	BB	38.13%	9.14%
			BB(CIoU)	39.06%	7.21%
		Center	BB+FV	72.58%	39.59%
			FV+CIoU	74.45%	43.97%
		Vertex	BB+FV	72.77%	39.67%
			FV+CIoU	74.48%	43.94%

The backbones comprise large VGGNet in SSD and small MobilenetV1 in SSDLite, and the input sizes consist of 300×300 and 512×512. The regression strategies include center-based, and vertex-based approaches, as shown in Fig. 5.3. If not specified, VGGNet, input size 512×512, and center-based regression strategy are adopted. Our proposed FV+CIoU can achieve the best F1-score for all settings, especially for the large IoU threshold of 0.9, which proves the proposed vertex adjustment loss is plug-and-play to improve vertex prediction.

5.3.5 License plate detection on CCPDv2 subsets

We compare our proposed FV+CIoU with BB, BB(CIoU), and BB+FV on the subsets of CCPDv2. As shown in Fig. 5.5, FV+CIoU can achieve the best F1-score on all the subsets, especially for the large IoU thresholds. BB and BB(CIoU) achieve the worst performance because it can only detect the horizontal bounding box of the license plate. Moreover, the license plates in the subsets Rotate and Tilt are highly tilted due to in-plane and out-of-plane rotation. BB and BB(CIoU) perform even worse on these subsets, and F1-score becomes zero when the IoU threshold is larger than 0.7. FV+CIoU can acquire more performance gain than other methods on these subsets because of more accurate vertex prediction.

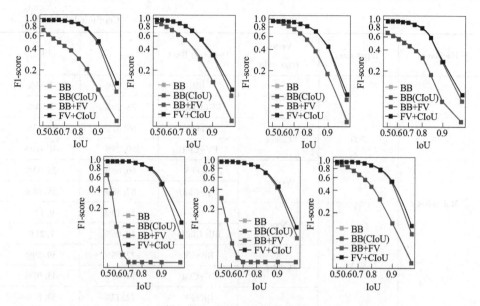

Fig. 5.5 Expermental result

5.3.6 License plate recognition on CCPDv2

As shown in Table 5.3, our proposed FV+CIoU achieves the best recognition accuracy on all the CCPDv2 subsets with different recognition metrics, closest to the accuracy of GT(Rectified). Especially for the subsets Rotate and Tilt with highly tilted license plates, FV+CIoU performs 55.93% and 45.15% higher than BB(CIoU). As can be concluded from Table 5.1, Table 5.2, Table 5.3 and Fig. 5.5, the recognition accuracy is highly correlated with the F1-score because a higher F1-score indicates more accurate multidirectional license detection, thus improving license plate recognition.

Table 5.3 License plate recognition on CCPDv2

Method	Avg	DB	Blur	FN	Rotate	Tilt	Challenge
GT	22.63%	18.10%	7.32%	30.29%	21.39%	13.16%	26.07%
GT(Rectified)	43.65%	27.99%	8.81%	49.98%	76.24%	62.08%	32.25%
BB	20.19%	16.64%	5.04%	25.56%	19.30%	12.97%	23.16%
BB(CIoU)	20.52%	16.97%	5.26%	25.95%	19.96%	13.04%	23.42%
BB+FV	40.35%	23.21%	5.75%	41.60%	72.77%	60.77%	28.58%
FV+CIoU(Ours)	42.38%	25.52%	6.68%	44.86%	75.89%	61.19%	31.42%

GT: The license plate is cropped based on the ground-truth horizontal bounding box.

GT(Rectified): The license plate is rectified horizontally based on the ground-truth vertices.

5.3 Experiments

Some qualitative examples are shown in Fig. 5.6. BB can accurately detect the horizontal bounding box. However, since the character features are extracted in a vertical slice, it will cause recognition errors due to the interference between characters and from background noises. Moreover, since the regression distance of each vertex differs, BB+FV can not accurately regress the distant vertices based on the regression base point, thus affecting character recognition at the edge of the license plate. Our proposed FV+CIoU can correctly recognize the multidirectional license plate based on the adjusted vertices.

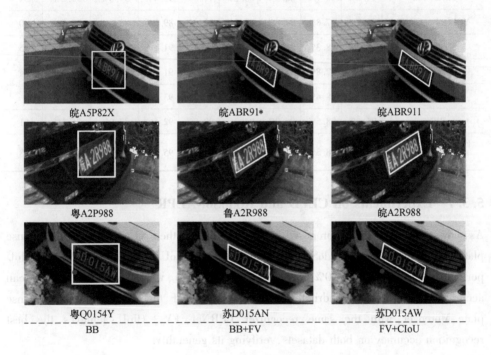

Fig. 5.6 Result analysis

As shown in Table 5.4, we compare the proposed method with other state-of-the-art methods on CCPDv1. We use the proposed detector FV + CIoU to detect the multidirectional license plate, then use the existing recognizer LPRNet to recognize the rectified license plate. The results prove our method can achieve the best average performance. Especially for the subsets Rotate and Tilt, we can achieve the best performance, proving the ability for multidirectional license plate detection and recognition.

Table 5.4 Comparative recognition experiments on CCPDv1

Method	Avg	Base	DB	FN	Rotate	Tilt	Weather	Challenge	FPS
Cascade classifier + HC	58.9%	69.7%	67.2%	69.7%	0.1%	3.1%	52.3%	30.9%	29
Faster-RCNN + HC	92.8%	97.2%	94.4%	90.9%	82.9%	87.3%	85.5%	76.3%	13
Zherzdev et al.	93.0%	97.8%	92.2%	91.9%	79.4%	85.8%	92.0%	69.8%	56
YOLOv2 + HC	93.7%	98.1%	96.0%	88.2%	84.5%	88.5%	87.0%	80.5%	36
TE2E	94.4%	97.8%	94.8%	94.5%	87.9%	92.1%	86.8%	81.2%	3
Silva et al.	94.6%	98.7%	86.5%	85.2%	94.5%	95.4%	94.8%	91.2%	31
YOLOv4 + CRNN	94.7%	97.8%	94.6%	87.3%	82.9%	89.9%	83.3%	75.7%	40
SSD300 + HC	95.2%	98.3%	96.6%	95.9%	88.4%	91.5%	87.3%	83.8%	35
SYOLOv4 + CRNN	95.3%	97.8%	95.0%	88.9%	84.9%	91.5%	90.4%	77.1%	34
Zhang et al.	95.4%	98.4%	97.0%	90.6%	92.7%	93.5%	86.9%	84.8%	—
RPNet	95.5%	98.5%	96.9%	94.3%	90.8%	92.5%	87.9%	85.1%	61
FV+CIoU(Ours) + LPRNet	95.9%	98.3%	95.2%	93.5%	94.6%	95.6%	91.1%	83.8%	38

5.3.7 Experiments on CLPD and RodoSol-ALPR

As shown in Table 5.5, our proposed FV+CIoU achieves the best F1-score on the license plate datasets CLPD and RodoSol-ALPR. For the large IoU threshold of 0.9, FV+CIoU performs 17.26% and 50.60% higher than BB(CIoU), which proves FV+CIoU can accurately predict the quadrilateral bounding box of the multidirectional license plate. Moreover, with the same recognizer LPRNet, FV+CIoU achieves the best recognition accuracy on both datasets, verifying its generality.

Table 5.5 License plate detection and recognition on CLPD and RodoSol-ALPR

Dataset	Method	Detection		Recognition
		$F1_{0.5:0.95}$	$F1_{0.9}$	Accuracy
CLPD	GT	—	—	60.83%
	GT(Rectified)	—	—	65.00%
	BB	63.29%	17.98%	56.94%
	BB(CIoU)	64.18%	18.98%	58.61%
	BB+FV	76.45%	34.43%	61.11%
	FV+CIoU(Ours)	78.78%	36.24%	64.17%

Continued Table 5.5

Dataset	Method	Detection		Recognition
		$F1_{0.5:0.95}$	$F1_{0.9}$	Accuracy
RodoSol-ALPR	GT	—	—	39.66%
	GT(Rectified)	—	—	40.75%
	BB	71.33%	35.50%	33.24%
	BB(CIoU)	71.72%	37.99%	34.23%
	BB+FV	89.70%	86.58%	37.23%
	FV+CIoU(Ours)	90.86%	88.59%	38.28%

5.3.8 Experiments on ICDAR2015

We use the official scripts to evaluate the text detection performance. As shown in Table 5.6, our proposed FV+CIoU achieves the best Precision, Recall, and F1-score on the scene text dataset ICDAR2015, which proves the proposed vertex adjustment loss also applies to skewed scene text detection.

Table 5.6 Text detection on ICDAR2015

Method	$Precision_{0.5}$	$Recall_{0.5}$	$F1_{0.5}$
BB	58.44%	40.49%	47.84%
BB(CIoU)	59.23%	40.49%	48.10%
BB+FV	68.72%	48.87%	57.12%
FV+CIoU(Ours)	71.32%	50.17%	58.90%

Please see colorful images

Chapter 6　Fast Recognition for Multidirectional and Multi-type License Plates with 2D Spatial Attention

The multi-type license plate can be roughly classified into two categories, i. e. , one-line and two-line. Many previous methods are proposed for horizontal one-line license plate recognition and consider license plate recognition as a one-dimensional sequence recognition problem. However, for multidirectional and two-line license plates, the features of adjacent characters may mix together when directly transforming a license plate image into a one-dimensional feature sequence. To solve this problem, we propose a two-dimensional spatial attention module to recognize license plates from a two-dimensional perspective. Specifically, we devise a lightweight and effective network for multidirectional and multi-type license plate recognition in the wild. The proposed network can work in parallel with a fast running speed because it does not contain any time-consuming recurrent structures. Extensive experiments on both public and private datasets verify that the proposed method out-performs state-of-the-art methods and achieves a real-time speed of 278 FPS.

6.1　Problem formulation

License plate recognition(LPR) can be applied to many real applications, such as traffic control, vehicle search, and toll station management. There are many types of license plates(LPs) in real scenarios. As shown in Fig. 6.1, multi-type LPs could be roughly classified into two categories, i. e. , one-line and two-line. Besides, license plates in the wild are multidirectional due to different shooting angles. However, many existing approaches are only aim at horizontal one-line license plate recognition. Therefore, it is still challenging to recognize multidirectional and multi-type license plates.

Many previous LPR methods consider license plate recognition as a one-dimensional sequence recognition problem. They transform input images into one-dimentional (1D) sequences by Convolutional Neural Networks (CNNs) for visual features extraction and utilize Connectionist Temporal Classification (CTC) loss or sequential attention mechanism to predict characters. However, these methods are not suitable for

Fig. 6.1 Examples of multi-type LPs with multiple directions

multidirectional and two-line license plate recognition. As shown in Fig. 6.2(a), the features of adjacent characters may mix together when directly transforming LP images into 1D features. Besides, many methods propose to rectify LP images or features to the horizontal direction by a rectification module for multidirectional license plate recognition. However, the rectification process has a large amount of calculation and is time-consuming. In addition, only a few methods are proposed for multi-type especially two-line license plate recognition. However, these methods are based on character segmentation, requiring lots of character location annotations and complicated post-processing.

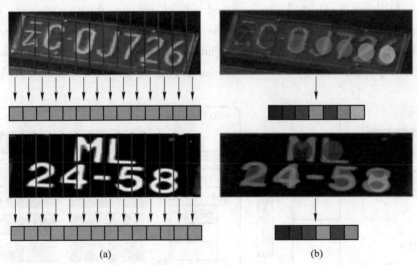

Fig. 6.2 Illustration of extracting visual features in the 1D or 2D space

We choose to use a two-dimensional (2D) attention mechanism to develop a fast and effective network for multidirectional and multi-type license plate recognition. Different from existing sequential attention algorithms on 2D visual feature, we consider LPR as an image classification problem that the prediction of each character in the LP is independent and parallel. Specifically, we propose a 2D spatial attention module without any recurrent structures to extract the most distinctive features for each character in the 2D space (see Fig. 6.2(b)). Based on this, we devise a fast and effective network for multidirec-tional and multi-type license plate recognition. In detail, the proposed network is composed of three components, i.e., a CNN-based encoder, a 2D spatial attention module, and a fully connected (FC) layer based decoder. Given an input image, the network first extracts visual features using the encoder, then aligns the whole LP features with each character by the 2D spatial attention module by reweighting features, and finally decodes attentioned features to predict characters.

6.2 Methodology

Our proposed method aims at recognizing multidirectional and multi-type license plates with a fast speed. Given an input image, the proposed network first extracts visual features of the whole LP and then calculates the attention weights of each character in the 2D space. According to the calculated attention weights, the network calculates the attentioned features which represent the most distinctive features for each character. Finally, the network decodes the attentioned features to characters in parallel. As shown in Fig. 6.3, our proposed network contains three components, i.e., the encoder, the 2D spatial attention module, and the decoder. "Conv" and "Deconv" stand for convolutional layer and deconvolution layer with kenerl size and channel. "Pool", "FC", and "Transpose" refer to max pooling layer with kenerl size, FC layer,

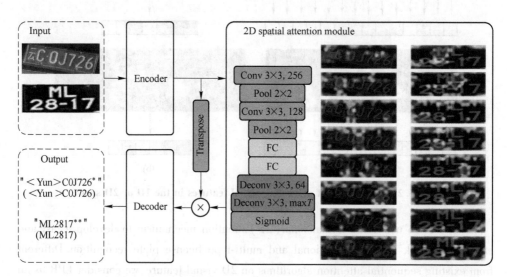

Fig. 6.3 Overall architecture of our proposed network

and matrix transpose, respectively. "$\max T$" is the max length of LP texts during training, and the operator of "$\otimes$" means matrix multiplication. "$*$" in the output results represents a blank character and it will be removed in the testing phase. The following sections will illustrate the details of these three components.

We adopt a lightweight backbone from Holistic CNN (HC) to extract visual features. Specifically, we abandon all the FC layers used as character classifiers and change the downsampling rate of features by alternating the number and the kernel size of pooling layers. Precisely, we empirically choose a downsampling rate r of 4, and the detailed structure of the encoder is shown in Table 6.1. The input is set to be gray images with the fixed size (W, H) of $(96, 32)$. The stride and the padding of each convolutional layer are 1. A batchnorm layer and a RELU layer are added to each convolutional layer.

Table 6.1 Description of the encoder

Name	Layer type	Channel	Size	Input	Output
Conv1-1	convolution	32	3 × 3	96 × 32 × 1	96 × 32 × 32
Conv1-2	convolution	32	3 × 3	96 × 32 × 32	96 × 32 × 32
Conv1-3	convolution	32	3 × 3	96 × 32 × 32	96 × 32 × 32
Conv2-1	convolution	64	3 × 3	96 × 32 × 32	96 × 32 × 64
Conv2-2	convolution	64	3 × 3	96 × 32 × 64	96 × 32 × 64
Conv2-3	convolution	64	3 × 3	96 × 32 × 64	96 × 32 × 64
Pool1	max pooling	64	2 × 2	96 × 32 × 64	48 × 16 × 64
Conv3-1	convolution	128	3 × 3	48 × 16 × 64	48 × 16 × 128
Conv3-2	convolution	128	3 × 3	48 × 16 × 128	48 × 16 × 128
Conv3-3	convolution	128	3 × 3	48 × 16 × 128	48 × 16 × 128
Pool2	max pooling	128	2 × 2	48 × 16 × 128	24 × 8 × 128

The features extracted by the encoder contain the information of the total characters in an LP image. We choose to utilize attention mechanism to obtain the most distinctive features for each character, which will learn the proper alignment of the whole LP features with each character.

The Sequential Attention Mechanism is proposed for sequence recognition problems. The most commonly used attention algorithms for LP and text recognition is based on Bahdanau et al's research. The attention weights of t time is calculated as Equation (6.1), where $f_h(\cdot)$, $f_a(\cdot)$, and $f_{\text{softmax}}(\cdot)$ represent an RNN cell, scoring function and softmax function, respectively. y_{t-1} is the output of the last time, and h_j represents the j-th hidden state of the input.

$$s_t = f_h(s_{t-1}, y_{t-1}, \sum_{j=1}^{T_x} \alpha_{tj} h_j)$$
$$\alpha_{tj} = f_{\text{softmax}}(f_a(s_{t-1}, h_j))$$
(6.1)

It can be found that attention weights are coupled with previous outputs. Therefore, when the previous character is wrongly predicted, the attention weights of the current time might be wrong. In addition, since there is little semantics between LP characters, the recurrent structure for modeling the relationship of adjacent characters is unnecessary. In view of the above two aspects, we consider LPR as an image classification task and propose a novel spatial attention module to predict each character in the LP independently and in parallel.

The proposed 2D spatial attention module aims to extract the most distinctive features for every character in the 2D space. As illustrated in Fig. 6.4, for every character in an LP image, its corresponding attentioned features are the weighted sum of the visual features according to the calculated attention weights. Attentioned features for each character are the weighted sum of the visual features according to the calculated attention weights. The different colors on attentioned features represent different character orders and the different colors on visual features represent different attention weights. "maxT" refers to the max length of LP texts in the training phase. Then, all the attentioned features are sent to the decoder in parallel by concatenating features together along the width dimension.

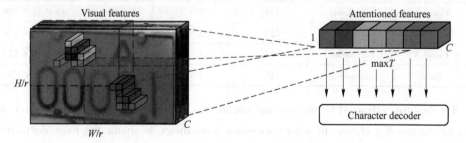

Fig. 6.4 The function sketch map of our proposed spatial attention algorithm

To ensure both global and local information could be covered, we adopt two kinds of operators, i.e., convolution (Conv) and FC. Specifically, the 2D spatial attention module takes the visual features $F \in R^{C \times (H/r \times W/r)}$ extracted from encoder as input and calculates attention weights $A \in R^{\text{max}T \times (H/r \times W/r)}$ as Equation (6.2).

$$A = f_{\text{Sig}}(f_{\text{Deconv}}(f_{\text{FC}}(f_{\text{Conv-Pool}}(F))))$$
(6.2)

where, C is the channel of visual features, maxT is the max length of LP texts in the dataset, $f_{\text{Sig}}(\cdot)$, $f_{\text{Deconv}}(\cdot)$, $f_{\text{FC}}(\cdot)$ and $f_{\text{Conv-Pool}}(\cdot)$ refers to sigmoid operator, deconvolution operator, fully connected operator, and convolution with max pooing operator. Detailed configuration is described in Fig. 6.3. In this way, the attention weights of different characters are calculated independently and in parallel. For the k-th character of the LP, its attention weight $A_k \in R^{H/r \times W/r}$ is the k-th slice of A in channel dimension, and its attentioned feature $F_{A(k)} \in R^{C \times 1 \times 1}$ is computed as Equation(6.3).

$$F_{A(k,j)} = \sum_{m,n=0}^{m=H/r,n=W/r} A_{k,m,n} \times F_{j,m,n}, j \in [0, C-1] \qquad (6.3)$$

To realize parallel computation of attentioned feature, we first reshape A to $A' \in R^{\max T \times (H/r \times W/r)}$, and F to $F' \in R^{C \times (H/r \times W/r)}$, then utilize matrix multiplication to calculate attentioned features for all the characters simultaneously (see Equation(6.4)), where F^T represents the transpose matrix of F.

$$F_A = A'(F')^T \qquad (6.4)$$

The decoder of the proposed method is a character classifier. Considering that there is little semantic relation between adjacent characters, we adopt one FC layer as the decoder for convenience. The cross-entropy loss is employed for training. Because the length of the LP text changes with the LP type, we increase a blank character to deal with variable lengths of LPs. For the LP with fewer characters than max$T$, blank characters are added to the end of ground truth until the length reaching max$T$ in the training phase.

6.3 Experiments

The existing public datasets are basically composed of one-line LPs. To demonstrate the effectiveness of our proposed method on multi-type LPs, including one-line and two-line LPs, we evaluate our method on both public datasets and private datasets which cover several types of LP.

SynthLP is generated according to the layouts of Chinese license plates in the way of using Python Image Library (PIL). Besides, we add random affine transformation, Gaussian noise, and RGB channel addition noise to the generated images. SynthLP covers two types of LP and 31 provinces in China (see Fig. 6.5(a)). Each type of LPs contains 80000 images. MBLP is collected by mobile phones in different cities of China. After careful annotation, a total of 26426 LP images are obtained. According to different layouts of LP, MBLP is categorized into three subsets (see Fig. 6.5(b)), i.e., one-line plates with a green background (l1-g), one-line plates with a yellow background (l1-y), and two-line plates with a yellow background (l2-y). The number of LPs of each subset is 8412, 5514, and 12500 sequentially. The ratio of training set to testing set is 9 : 1. DRLP is collected by drive recorders in different cities of

China. After careful annotation, a total of 50536 LP images are obtained. According to different layouts of LP, DRLP is classified into five subsets (see Fig. 6.5(c)), i.e., one-line plates with a blue background (l1-b), two-line plates with a yellow background(l2-y), two-line plates with a black background(l2-bl), one-line plates with a black background containing 6 characters (l1-bl6), and one-line plates with a black background containing 7 characters (l1-bl7). The number of LPs of each subset is 33980, 2946, 736, 5791, and 7083 sequentially. The ratio of training set to testing set is 9 : 1. CCPD is currently the largest public LP dataset. It totally contains 8 subsets divided by the complexity and photographing condition of LP images, such as tilt, weather and so on. Among them, 10k images are used for training, and 15k images are used for testing. CDLP is collected from the internet and only for testing. It contains 1200 images covering different photographing conditions, vehicle types, and region codes of Chinese provinces.

Fig. 6.5 Examples of private LP datasets

Performances on MBLP and DRLP: We implemented the three most commonly used text recognition algorithms for LPR, i.e., CTC-based, 1D attention-based, and 2D attention-based. Except for the difference of some convolutional kernel size, the encoder networks of these implementations are the same as ours. The results are shown in Table 6.2.

Table 6.2 License plate recognition accuracy on MBLP and DRLP

Datasets	Dim	All	MBLP			DRLP				
Methods (images)		7700	l1-g	l1-y	l2-y	l1-b	l2-y	l2-b	l1-b6	l1-b7
			842%	552%	1250%	3398%	295%	74%	580%	709%
CTC	1D	83.54%	91.69%	86.77%	84.56%	78.25%	75.25%	63.51%	90.00%	95.35%
Atten1D	1D	83.92%	91.81%	87.14%	85.76%	78.46%	75.59%	56.76%	91.03%	95.49%
Atten2D	2D	89.71%	**96.31%**	**94.93%**	91.12%	85.13%	79.32%	83.78%	**95.17%**	97.74%
HC ‡	2D	87.94%	94.18%	90.04%	90.08%	84.14%	**80.68%**	78.54%	91.21%	95.63%
Ours	2D	**90.23%**	96.19%	**94.93%**	**92.72%**	**85.63%**	80.33%	**85.13%**	95.00%	**97.88%**

The subsets represent different types of LPs. ‡ means that it is implemented by ourselves. Comparing the results of 1D-based methods with 2D-based methods, it shows that the 2D-based method has significant superiority over the 1D-based methods, which reveals that decoding on 2D features is more suitable for LPR. Furthermore, among the three 2D-based methods, our proposed method performs better in recognizing multi-type LPs, especially for two-line LPs.

Performances on CCPD: As illustrated in Table 6.3, † means that LP images used for testing are cropped by the ground truths. "Rot.", "Wea.", and "Cha." represent "Rotate", "Weather", and "Challenge", respectively. Our approach achieves the best performance, especially on non-horizontal subsets, such as "Rotate", "Tilt". Meanwhile, the proposed method achieves a fast inference speed of 3.6 ms per image, i.e., 278 FPS. Due to the difference between annotated and detected bounding boxes, the LP images cropped by these two kinds of bounding boxes is different. The performance on the two kinds of LP images fluctuates slightly, which indicates our proposed method is robust to the variation of LP boundaries.

Table 6.3 License plate recognition accuracy on CCPD

Methods (images)	All	Base 100k	DB 20k	FN 20k	Rot. 10k	Tilt 10k	Wea. 10k	Cha. 10k	Time (ms)
SSD+HC	95.2%	98.3%	96.6%	95.9%	88.4%	91.5%	87.3%	83.8%	25.6
TE2E	94.4%	97.8%	94.8%	94.5%	87.9%	92.1%	86.8%	81.2%	310
PRnet	95.5%	98.5%	96.9%	94.3%	90.8%	92.5%	87.9%	85.1%	11.7
MORAN	98.3%	99.5%	98.1%	98.6%	98.1%	98.6%	97.6%	86.5%	18.2
DAN	96.6%	98.9%	96.1%	96.4%	91.9%	93.7%	95.4%	83.1%	19.3
Zhang et al.	98.5%	99.6%	98.8%	98.8%	96.4%	97.6%	98.5%	88.9%	24.9
Ours	**98.74%**	**99.73%**	**99.05%**	99.23%	97.62%	**98.4%**	**98.89%**	88.51%	**3.6**
Ours †	98.73%	99.67%	99.04%	**99.24%**	**97.66%**	98.26%	98.83%	**89.17%**	

Performances on CDLP: Table 6.4 shows our method significantly outperforms the other methods. It indicates that our approach could extract the most distinctive feature for each character and has good generalization for different photographing conditions. Besides, there is almost one kind of region code of LPs in CCPD, whereas CDLP covers many region codes. Therefore, the accuracy with region code is relatively low.

Table 6.4 License plate recognition accuracy on CDLP

Methods	CDLP	
Criterion	ACC	ACC w/o RC
PRnet	66.5%	78.9%
Zhang et al.	78.9%	86.1%
Ours	80.3%	91.7%

To further demonstrate the effectiveness of our proposed 2D spatial attention algorithm on multidirectional and multi-type LPs, we conduct ablation experiments on MBLP, DRLP, and CCPD. "Rotate" and "Tilt" in CCPD are combined as the non-horizontal subset. Moreover, we divide the combination of MBLP and DRLP into the one-line subset and the two-line subset according to the number of LP text lines.

The proposed attention module contains two different types of operators, namely local and global operators. We first learn the influence of different types of operators. Besides, we remove the attention module and add several convolutional layers to validate the effectiveness of the attention module. In the way of adjusting the hyperparameters of the operators, such as kernel size, we ensure that different models have a comparable quantity of parameters. As shown in Table 6.5 and Fig. 6.6, the proposed attention module could boost accuracy on both non-horizontal and multi-type LP datasets. Besides, the attention module using both global and local operators is better than using only one of them.

Table 6.5 Ablation study on the 2D spatial attention modules with local or global operators

Methods	Local	Global	CCPD		MBLP&DRLP		
			All	Non-hor.	All	One-line	Two-line
Ours(w/o. Atten)	√		98.53%	94.33%	87.96%	88.29%	86.72%
Ours(w/o. FC)			98.63%	97.61%	88.66%	88.73%	88.38%
Ours(w/o. Conv)	√	√	98.66%	97.74%	89.19%	89.01%	89.87%
Ours		√	98.73%	97.96%	90.23%	92.26%	90.12%

Furthermore, the suitable number of operators for the 2D spatial attention module is learned. We implement three spatial attention modules with different numbers of Conv, FC, and Deconv layers. Table 6.6 indicates that the performance grows with the increase of operators and tends to be saturated when the number of each operator reaching 2.

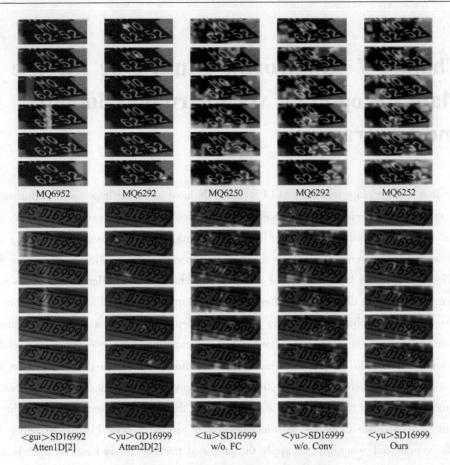

Fig. 6.6 Examples of attention visualization

Table 6.6 Ablation study on the 2D spatial attention modules with different numbers of operators

Methods	Configuration	CCPD		MBLP&DRLP		
		All	Non-hor.	All	One-line	Two-line
Ours-1	Conv × 1-FC×1-Deconv × 1	98.04%	96.48%	87.05%	87.46%	85.48%
Ours-2	Conv × 2-FC×2-Deconv × 2	98.73%	97.96%	90.23%	92.26%	90.12%
Ours-3	Conv × 3-FC× 3-Deconv × 3	98.65%	97.72%	90.15%	90.10%	90.36%

Please see colorful images

Chapter 7　Improving Multi-type License Plate Recognition via Learning Globally and Contrastively

Previous license plate recognition(LPR)methods have achieved impressive performance on single-type license plates. However, multi-type license plate recognition is still challenging due to various character layouts and fonts. There are two main problems: one is that recognition models are prone to incorrectly perceive the location of characters due to diverse character layouts, and the other is that characters of different categories may have similar glyphs due to various fonts, causing character misidentification. Therefore, to solve the above problems, we propose two plug-and-play modules based on an attention-based framework for multi-type license plate recognition. First, we propose a global modeling module to integrate character layout information to precisely perceive the location of characters, thus generating accurate predictions. Second, a position-aware contrastive learning module is proposed to enhance the robustness and discriminability of features to alleviate character misidentification of similar glyphs. Finally, to verify the effectiveness and generality, we apply the proposed modules to six baseline models, and the results demonstrate that the proposed method can achieve state-of-the-art performance on three multi-type license plate datasets. Moreover, extensive experiments prove that our proposed modules can significantly improve performance by 6.8% on RODOSOL-ALPR with a small parameter increase.

7.1　Problem formulation

License plate recognition (LPR) is integral to intelligent transport systems, and it has numerous practical applications in traffic management, law enforcement, and parking systems. Many previous LPR approaches focus on single-type license plates and achieve promising results. However, multi-type license plate recognition remains challenging due to diverse character layouts and fonts.

To recognize license plates with various character layouts, some LPR methods propose first to detect or segment characters and then identify them. However, these methods require character-level annotations for training and prediction assembly for post-processing, which is labor-intensive and costly. Therefore, many LPR methods consider the license plate as a sequence and leverage attention-based methods to extract character features in the one-dimensional (1D) space. However, the characters of two-line license plates are arranged in two horizontal rows, and forcibly processing license plates in the 1D space will mix the features of the upper and lower rows, leading to poor performance. Therefore, some LPR methods propose 2D-attention algorithms to extract character features in the 2D space and then identify each character. We summarize these methods into three categories according to the attention algorithms and build three kinds of 2D attention-based LPR models to analyze the effectiveness of these methods in multi-type license plate recognition. These LPR models have the same architecture except for different attention approaches. However, as shown in Fig. 7.1(a), these methods suffer from attention misalignment, which is caused by the extracted visual features lacking global information about character layout to guide attention generation, leading to wrong predictions.

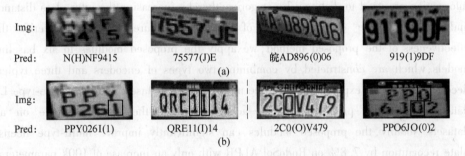

Fig. 7.1 The challenges of multi-type license plate recognition

Additionally, characters of different categories may have similar glyphs due to diverse fonts, leading to character misidentification. As shown in Fig. 7.1(b), "1" of the first LP is more similar to "I" than "1" of the second LP, and "O" of the third LP is more similar to "0" than "O" of the fourth LP. Therefore, learning a discriminative representation of characters is necessary. Currently, contrastive learning has shown superiority in representation learning in a variety of domains such as speech processing, language processing, and computer vision. However, generic contrastive learning methods in computer vision are designed for object recognition, which cannot be applied to sequence recognition such as license plate recognition. Hence, many text recognition methods propose to utilize instance-mapping functions or attention modules to yield the features of separate characters for contrastive learning. However, these methods neglect the position of characters, which is beneficial to distinguish characters of similar glyphs. Considering that the range of character categories for each position is fixed within the same license plate type due to the corresponding template, incorporating position information could enhance recognition. For example, the seventh character of a Brazilian license plate (the first LP in Fig. 7.1(b)) can only be a number, and the 3rd character of a California license plate (the third LP in Fig. 7.1(b)) can only be a letter.

To solve the above problems, we propose two plug-and-play modules based on an attention-based framework to improve multi-type license plate recognition. First, we propose a global modeling module to mine the relationships between characters. This way, the character layout information can be integrated into recognition models to enable the attention module to accurately perceive character locations, thus improving recognition performance. Second, to alleviate character misidentification of similar glyphs, we propose a position-aware contrastive learning module by pulling the character features of the same category and position together and pushing apart other character features in the representation space. This way, both visual features and position information are used to distinguish categories, thereby increasing the inter-class distance and enhancing the discriminability of character features. Finally, to verify the effectiveness of the proposed method, we apply the proposed modules to six baseline models, which are constructed by combining two types of encoders and three typical decoders. Extensive experiments on three multi-type LP datasets and two single-type LP datasets prove the proposed method achieves state-of-the-art performance on all datasets. Notably, the proposed modules can significantly improve multi-type license plate recognition by 7.8% on RodoSol-ALPR with only an increase of 100k parameters.

7.2 Methodology

As illustrated in Fig. 7.2, the proposed method follows an attention-based encoder-decoder framework. The global modeling module mines the relationship of characters to introduce character layouts for attention generation. The position-aware contrastive learning module contrasts character features in different positions and categories to enhance the robustness and distinction of character features. Our main contributions lie in two plug-and-play modules: global modeling module (GMM) and position-aware contrastive learning (PCL). We select six different models as baselines to verify their effectiveness and generalization, which are the combination of two encoders with different parametric quantities and three typical attention-based decoders. The details of each module are described below.

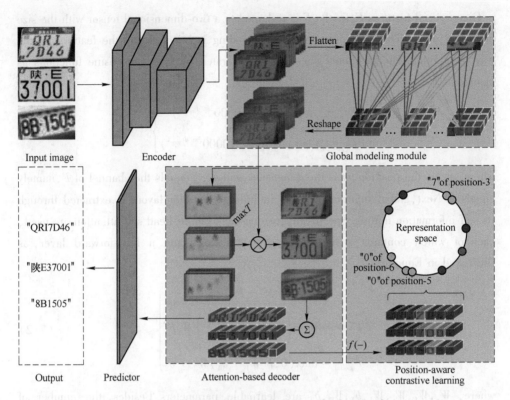

Fig. 7.2 The architecture of the proposed network

7.2.1 Encoder

We selected a lightweight backbone from Holistic CNN(HC) as the base encoder. Moreover, to further decrease the model parameters and computational cost, we remove one-third of the layers of the base encoder and reduce channels to half of the original as the small encoder. For convenience, the features extracted by the encoder are denoted as $F \in \mathbb{R}^{C \times (H \times W)}$, where C, H, W represent the channel, the width, and height of the features, respectively. The input size is set to (32,96), and the output size is (8,24).

7.2.2 Global modeling module

Diverse character layouts pose a challenge for the subsequent decoder to generate accurate attention. Therefore, we propose a global modeling module to introduce character layout information for the decoder by mining the relationships of characters. Inspired by the self-attention algorithm, we designed the global modeling module as follows.

Given visual features F, firstly, F is reshaped to a two-dimensional tensor with the size of (C, (H, W)). Then, absolute positional encoding is injected into the features in the way of addition. The positional encoding is generated by sine and cosine functions as Equation(7.1).

$$\left.\begin{array}{l} PE_{(pos,2m)} = \sin(pos/10000^{2m/d_{model}}) \\ PE_{(pos,2m+1)} = \cos(pos/10000^{2m/d_{model}}) \end{array}\right\} \quad (7.1)$$

where, pos is the position, m is the dimension, and d_{model} equals the channel of F, namely equals C. Next, global information F_{global} including character layout is extracted through its self-information mining. Specifically, two cascaded multi-head self-attention are used, each of which contains a multi-head attention layer and a feed-forward layer, as illustrated in Equation(7.2).

$$\left.\begin{array}{l} F' = F + PE \\ F'' = \text{softmax}\left(\dfrac{W_qF'(W_kF')^T}{\sqrt{d_{model}}}\right)W_vF' \\ F_{global} = \max(0, F''W_1 + b_1)W_2 + b_2 \end{array}\right\} \quad (7.2)$$

where, $W_q, W_k, W_v, W_1, b_1, W_2, b_2$ are learnable parameters. Besides, the number of heads is experimentally set to 2 and 4 for the small and base encoder, respectively. The inner model dimension is set to four times of d_{model}.

7.2.3 Decoder

The attention-based decoder aims to extract the representative features of each character from the visual features. Three typical 2D attention methods are adopted to verify the effectiveness of the proposed method fully. As shown in Fig. 7.3, the attention module can be considered as mapping a query and a key-value set by the similarity of the query and key, and then aggregating the value according to the similarity. The key-value set is usually derived from visual features while the query is generated differently referring to the specific attention.

(1) Parallel visual attention module (PVAM): PVAM is proposed by semantic reasoning network(SRN). As shown in Fig. 7.3(a), PVAM proposes to use the character reading order as the query to realize parallel attention. The key-value set (k_{ij}, v_{ij}) is derived from 2D visual features F_{global}, and the query derived from reading order embedding(O_t), the algorithm of PVAM can be summarized as Equation(7.3).

7.2 Methodology

$$\left.\begin{array}{c} k_{i,j} = v_{i,j} \in F_{\text{global}} \\ e_{t,ij} = W_e^T \tanh(W_o O_t + W_k k_{ij}) \\ \alpha_{t,ij} = \dfrac{\exp(e_{t,ij})}{\sum\limits_{\forall i,j} \exp(e_{t,ij})} \\ g_t = \sum\limits_{\forall i,j} \alpha_{t,ij} v_{ij} \end{array}\right\} \quad (7.3)$$

where, W_e, W_o, W_v are trainable weights, $\tanh(\ )$ is tangent activation function, i,j,t are the indices of width, height and reading order, respectively. g_t is the attended features, representing the character features of t-th order.

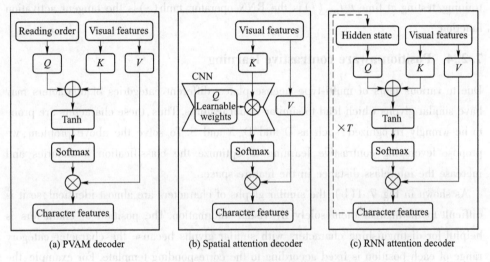

(a) PVAM decoder   (b) Spatial attention decoder   (c) RNN attention decoder

Fig. 7.3 Schematic diagram of different decoders

(2) Spatial Attention (S-A): As shown in Fig. 7.3(b), the spatial attention-based decoders utilize convolution neural networks (CNNs) to calculate attention weights of each character in the spatial dimension ($W \times H$ dimension). Considering simplicity and inference speed, we chose SALPR as one of the baseline decoders. SALPR adopts a U-Net-like network for attention calculation. The dimension of attention is max $T \times H \times W$, where max$T$ is the max length of all license plates in the dataset. Each channel represents the spatial attention of the corresponding character. For example, channel 0 is the spatial attention of the first character.

(3) RNN-based Attention (RNN-A): As shown in Fig. 7.3(c), the RNN-based attention decoder is very similar to PVAM, except that the key is derived from the hidden state of the RNNs. Denoting current hidden state as h_t, the RNN-based attention algorithm can be

summarized as Equation(7.4).

$$\left.\begin{aligned}
k_{i,j} &= v_{i,j} \in F_{\text{global}} \\
e_{t,ij} &= W_e^T \tanh(W_h h_{t-1} + W_k k_{ij}) \\
\alpha_{t,ij} &= \frac{\exp(e_{t,ij})}{\sum_{\forall i,j} \exp(e_{t,ij})} \\
g_t &= \sum_{\forall i,j} \alpha_{t,ij} v_{ij} \\
h_t &= f_{\text{RNN}}([g_t, y_t], h_{t-1})
\end{aligned}\right\} \quad (7.4)$$

where, W_e, W_h, W_k are trainable weights, y_t is the ground truth/predicted result during training/testing at time t, $f_{\text{RNN}}(\cdot)$ is the RNN operator, $\tanh(\cdot)$ is the tangent activation function.

7.2.4 Position-aware contrastive learning

Due to various fonts of multi-type license plates, different categories of characters may have similar glyphs which lead to similar visual features. Thus, these characters are prone to be wrongly recognized, such as 0 and D, 5 and S. To solve the above problem, we propose leveraging contrastive learning to optimize the classification boundaries and increase the interclass distance in the feature space.

As shown in Fig. 7.1(b), the similar glyphs of characters are almost identical, so it is difficult to distinguish them solely by visual information. The position of characters is helpful for distinguishing characters with similar glyphs because the character category range of each position is fixed according to the corresponding template. For example, the first three characters of a Brazilian motorcycle LP are letters, and the last four are numbers. Therefore, we propose position-aware contrastive learning to take advantage of both visual and position information to enhance the distinction of different character categories in the feature space.

As shown in Fig. 7.4, position-aware contrastive learning can be described as follows. Given the features of a character as an anchor sample, the features of characters with the same category and position are regarded as positive samples, and other features, including characters with the same category but different positions, are negative samples. Then, the anchor and positive samples are pulled together and negative samples are pushed away from the anchor in the representation space, optimizing the classification boundaries. Specifically, since a license plate contains a series of characters, the features of single character $G(g_t \in G)$ obtained by the decoder are used for contrastive learning. Then, a multi-layer perception(MLP), including two fully connected layers and an L2 normalization layer, is adopted as the projection network to map features G into comparable representations Z. Besides, the cosine distance is used to measure the distance between features. The position-aware contrastive learning loss L_{con} is defined as Equation(7.5).

$$L_{PCL} = \sum_{i=0}^{N} \frac{-1}{P(i)} \sum_{p \in P(i)} \lg \frac{\exp(z_i \cdot z_p/\tau)}{\sum_{k \in K(i)} \exp(z_i \cdot z_k/\tau)} \qquad (7.5)$$

where, $z_i \in Z(i = 0, 1, \cdots, N)$, N is the total number of characters in a mini-batch. $P(i)$ is the set of positive samples and $K(i)$ is other samples except z_i in the same batch. τ is a scalar temperature parameter, set to 0.1.

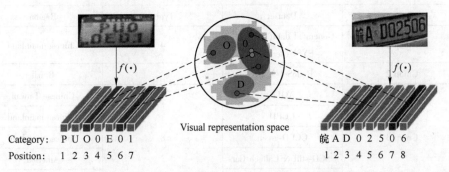

Fig. 7.4　Position-aware contrastive learning

7.2.5　Prediction module

The prediction module maps character features to class probability. Specifically, a fully connected layer is adopted as the prediction module. Besides, blank characters are appended to the end of the license plate strings to ensure uniform character length across all license plates, facilitating parallel training. The cross-entropy loss is used as recognition loss.

$$L_{recog} = -\sum_{i=0}^{C_{class}} y_i \lg(p_i) \qquad (7.6)$$

where, C_{class} is the number of character categories (including blank label), y_i equals 1 when i-th class is the ground truth, otherwise y_i is 0, p_i is the probability of i-th class. Therefore, the total loss of the model L is defined as Equation(7.7).

$$L = L_{recog} + \lambda L_{PCL} \qquad (7.7)$$

λ is experimentally set to 0.1.

7.3 Experiments

7.3.1 Datasets

To verify the effectiveness and generality of the proposed method, we use three multi-type license plate datasets and two commonly used single-type license plate datasets.

(1) Multi-type License Plate Datasets: As shown in Table 7.1, three groups of experimental datasets are integrated according to the number of types and the domain differences between the training and test data.

Table 7.1 The description of datasets used in the experiments

	Multi-type license plate		
Group	Dataset	Types	Region
Group 1	Generated data(training) SYSU(testing)	4	Chinese mainland
Group 2	RodoSol-ALPR	4	Brazil
Group 3	AOLP	8	Chinese Taiwan
	CCPD		Chinese mainland
	CCPD-green		Chinese mainland
	USCD-still & Caltech Cars		America
	RodoSol-ALPR		Brazil

Single-type license plate		
Dataset	Types	Region
CCPD	1	Chinese mainland
AOLP	1	Chinese Taiwan

Group 1 dataset is the license plates collected in mainland China. SYSU is a Chinese multi-type license plate dataset collected under surveillance. As shown in Fig. 7.5(a), four types of license plates are selected, namely one-line blue (line1 blue), one-line white (line1 white), one-line green (line1 green), and two-line yellow (line2 yellow) license plates. Due to the small amount of data, the SYSU is only used for testing. To train recognition models, 2500 license plates of each type were generated following the corresponding templates, as depicted in Fig. 7.5(a). In addition, random blur noise, random brightness perturbation, random affine transformation, and random background are added to generated images.

Group 2 dataset is the license plates collected in Brazil. RodoSol-ALPR is captured by static cameras day and night. As shown in Fig. 7.5(b), it contains four license plate types, each containing 5000 images. The ratio of training, validation, and testing is 2:1:2.

Group 3 dataset is license plates in multiple regions, including eight types of license plates from six public datasets, namely, AOLP, CCPD, CCPD-green, USCD-still, Caltech Cars, and RodoSol-ALPR. As shown in Fig. 7.5(c), AOLP consists of Chinese Taiwan LPs, CCPD, and CCPD-green contain two kinds of LPs in mainland China. USCD-Still and Caltech Cars are both American license plates. Since both have relatively few images, we combined them into one American LP sub-dataset, denoted as "US". RodoSol-ALPR consists of four kinds of Brazilian license plates. To balance the number of license plates of each type, 2000 of each type are taken for training. For CCPD and CCPD-green, we followed the original split protocol and randomly sampled 2000 images from the training dataset for training and 2000 images from the test dataset for testing. For AOLP, American LP (USCD-Still and Caltech Cars), we split the datasets following the protocol proposed by Rayson et al. Since their training dataset has less than 2000 images, we conducted character permutation and random image augmentation as Rayson et al. to expand the training data to 2000.

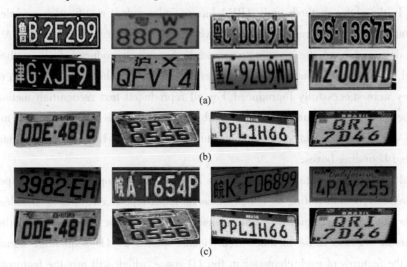

Fig. 7.5 License plate example

(2) Single-type License Plate Datasets: CCPD is currently the largest publicly available license plate dataset. CCPD contains eight subsets collected by mobile phones in mainland China with different photographing conditions. The 100000 images from the base subset are used for training and the rest of images is used for testing. AOLP consists

of 2049 images collected in Chinese Taiwan. It has three subsets, each of which is collected under different photographing conditions, i. e. , Access Control(AC), Traffic Law Enforcement(LE) and 374 Road Patrol(RP). Following previous work, we use two subsets for training and the remaining subset for testing.

7.3.2 Experiments on benchmarks

(1) Experiments on Multi-type License Plate Dataset: Since the test protocols for multi-type license plate recognition are defined by us, all comparative methods were reproduced to obtain the performance on the groups 1-3 datasets. Specifically, we reproduced two typical LPR methods, i. e. , HC which uses multiple classifiers to recognize the character at different locations, and LPRNet which is based on CTC. Besides, three recent LPR methods, which are improvements of typical methods, were reproduced to demonstrate the advantages of the proposed method. Among them, the method proposes three different sizes of recognition networks, and we compare the best-performing network, i. e. the large size model. Furthermore, since license plate recognition is a specific task in the text recognition domain, we also compared the proposed method with two commonly used text recognition methods, namely CRNN and TRBA. In the above methods, only TRBA has an independent rectification module. For a fair comparison, we also reproduced RBA, which has the same structure as TRBA except that it does not have the rectification module. For all reproduced LPR methods, we ensured that their performance on CCPD was consistent with the original work to confirm that they were successfully reproduced. For all reproduced text recognition methods, we used the official open-source codes. To study the effectiveness of the proposed method, we did not use any data augmentation during training and cropped LP images by annotated bounding boxes.

Results on group 1 Dataset: As shown in Table 7.2, our proposed method outperforms other methods, demonstrating its effectiveness on multi-type license plates. Besides, other methods can barely recognize two-line license plates except for the proposed method and HC. This is because these methods treat license plates as 1D sequential signals and extract the features of each character in the 1D space, which will mix the features of the top and bottom row of two-line LPs. Notably, even the proposed model with the small encoder can exceed many other methods, proving the superiority of the proposed method. To summarize, our proposed methods have the best performance and fast inference speed, suitable for practical license plate recognition applications.

7.3 Experiments

Table 7.2 Performance of group 1 dataset

Method		Group 1 dataset					Time (ms)
		Avg(2831)	Line1 blue (2577)	Line2 yellow (95)	Line1 green (101)	Line1 white (58)	
HC		40.8%	40.7%	47.4%	41.6%	36.2%	6.6
LPRNet		50.4%	55.4%	0.0%	10.9%	0.0%	3.8
SALPR		73.0%	73.3%	73.7%	67.3%	74.1%	4.0
Fan et al.		75.8%	76.9%	71.6%	51.5%	82.8%	4.0
Ke et al. (L)		74.2%	76.2%	1.1%	82.2%	61.7%	13.1
CRNN		79.4%	79.3%	1.0%	76.2%	75.9%	4.2
TRBA		89.7%	92.7%	1.1%	**95.1%**	96.6%	31.5
RBA		73.5%	75.5%	0.0%	85.2%	84.5%	29.7
Ours small	PVAM	85.3%	85.0%	88.4%	89.1%	91.4%	5.5
	S-A	85.8%	85.3%	84.2%	94.1%	**98.3%**	8.2
	RNN-A	74.6%	73.7%	75.8%	94.1%	81.0%	11.4
Ours base	PVAM	**94.3%**	**94.6%**	**90.5%**	94.1%	91.4%	6.5
	S-A	94.0%	94.2%	**90.5%**	92.1%	**98.3%**	10.1
	RNN-A	79.6%	78.5%	87.4%	93.1%	96.6%	12.8

Results on group 2 Dataset: As shown in Table 7.3, our proposed method is significantly better than the other methods on each subset. Notably, contrary to the results on group 1 dataset, the attention-based methods (TRBA and RBA) can correctly identify more than half of the two-line LPs (motor-me/br) in group 2. We think it is because these methods can forcibly fit two-line LPs with sufficient in-domain training data. Besides, CTC-based methods (CRNN and LPRNet) still have difficulty in correctly recognizing two-line LPs, and the mix of various types during training makes the performance on single-line LPs not high either. In conclusion, the proposed method is superior in multi-type license plate recognition.

Results on group 3 Dataset: As shown in Table 7.4, compared to the results on the group 2 dataset, though the performance of other models on the corresponding subsets improves as training data increases, the proposed method still achieves the best performance. Besides, it can be observed that the proposed method with a small encoder can also exceed other methods except for TRBA, which has an additional rectification module, and performance will be boosted if the rectification module is added.

Table 7.3 Performance of group 2 dataset

Method		Group 2 dataset					Time (ms)
		Avg(8k)	Cars-br (2k)	Cars-me (2k)	Motor-br (2k)	Motor-me (2k)	
HC		78.5%	83.6%	91.2%	69.6%	69.7%	6.6
LPRNet		43.3%	80.1%	89.1%	2.3%	1.7%	3.8
SALPR		92.3%	94.0%	96.5%	88.8%	89.9%	4.0
Fan et al.		90.1%	91.7%	94.9%	86.8%	87.1%	4.0
Ke et al. (L)		47.0%	83.2%	89.6%	7.2%	8.0%	13.1
CRNN		45.4%	73.8%	78.5%	9.4%	19.8%	4.2
TRBA		84.7%	95.4%	98.3%	60.1%	84.9%	31.5
RBA		78.8%	91.7%	96.7%	66.8%	60.0%	29.7
Ours small	PVAM	95.3%	96.5%	98.6%	91.6%	94.4%	5.5
	S-A	93.9%	95.8%	98.2%	89.4%	92.2%	8.2
	RNN-A	93.9%	95.4%	97.8%	89.8%	92.5%	11.4
Ours base	PVAM	**96.9%**	98.1%	**99.3%**	**94.0%**	**96.2%**	6.5
	S-A	96.6%	**98.2%**	99.0%	**94.0%**	95.4%	10.1
	RNN-A	96.2%	97.4%	98.7%	93.4%	95.4%	12.8

Table 7.4 The performance of group 3 dataset

Method		Avg	AOLP	CCPD	CCPD green	US	Cars-br	Cars-me	Motor-br	Motor-me
HC		64.7%	59.6%	37.8%	57.1%	34.0%	83.0%	90.2%	62.5%	61.1%
LPRNet		46.6%	44.5%	50.1%	55.3%	67.0%	84.5%	87.8%	0.9%	0.6%
SALPR		89.6%	93.0%	86.3%	77.0%	79.3%	96.0%	97.5%	87.9%	92.2%
Fan et al.		81.0%	84.2%	61.7%	67.0%	39.6%	93.4%	95.2%	85.2%	84.9%
Ke et al. (L)		58.1%	90.0%	77.5%	67.6%	75.5%	89.8%	91.0%	5.0%	6.2%
CRNN		72.1%	92.8%	81.3%	73.4%	81.1%	94.5%	97.0%	35.2%	43.8%
TRBA		92.1%	98.4%	93.1%	81.6%	84.0%	96.6%	98.4%	87.4%	93.8%
RBA		84.2%	95.0%	86.8%	78.4%	82.1%	95.6%	97.4%	74.2%	69.5%
Ours small	PVAM	92.5%	97.5%	91.3%	81.3%	86.8%	96.7%	98.2%	92.3%	93.7%
	S-A	92.7%	98.7%	92.7%	81.9%	86.8%	97.2%	98.3%	90.2%	94.1%
	RNN-A	91.4%	96.5%	89.6%	79.5%	84.9%	96.8%	97.5%	91.7%	91.7%

Continued Table 7.4

Method		Avg	AOLP	CCPD	CCPD green	US	Cars-br	Cars-me	Motor-br	Motor-me
Ours base	PVAM	94.8%	98.4%	94.7%	84.8%	88.7%	98.1%	98.7%	95.1%	96.7%
	S-A	95.2%	99.0%	95.2%	86.8%	90.6%	97.7%	98.6%	94.8%	97.1%
	RNN-A	94.2%	98.1%	94.4%	83.7%	87.7%	98.2%	98.1%	94.4%	95.4%

(2) Experiments on Single-type License Plate Dataset: To further demonstrate the superiority of the proposed method, we conduct experiments on several single-type LP datasets. We used data augmentation following the previous method when training on CCPD and AOLP. To fairly compare with other methods, we use YOLOv3 trained on the corresponding training sets to obtain LP bounding boxes.

Results on CCPD: CCPD is one of the most commonly used license plate datasets with the largest data volume among public LP datasets. As shown in Table 7.5, the proposed method performs best on almost all subsets. Notably, the proposed method has a significant advantage on the subset of tilted license plates (Rot. and Tilt). This is because the proposed method can effectively perceive character layouts and thus correctly recognize license plates with non-horizontal character arrangements.

Table 7.5 The performance of CCPD

Method		Avg	Base	DB	FN	Rot.	Tilt	Wea.	Cha
Xu et al.		95.5%	98.5%	96.9%	94.3%	90.8%	92.5%	87.9%	85.1%
Li et al.		94.4%	97.8%	94.8%	94.5%	87.9%	92.1%	86.8%	81.2%
LPRNet		93.0%	97.8%	92.2%	91.9%	79.4%	85.8%	92.0%	69.8%
DAN		96.6%	98.9%	96.1%	96.4%	91.9%	93.7%	95.4%	83.1%
Zhang et al.		98.5%	99.6%	98.8%	98.8%	96.4%	83.1%	98.5%	88.9%
SALPR		98.7%	99.7%	99.0%	99.2%	97.7%	98.3%	98.8%	89.2%
Zou et al.		97.5%	99.2%	98.1%	98.5%	90.3%	95.2%	97.8%	86.2%
Ours small	PVAM	98.6%	99.7%	98.8%	99.1%	97.5%	98.0%	98.5%	87.9%
	S-A	98.6%	99.7%	99.0%	99.1%	97.5%	98.2%	98.4%	88.4%
	RNN-A	98.0%	99.5%	98.3%	98.6%	96.0%	97.1%	97.7%	85.4%
Ours base	PVAM	**98.8%**	**99.7%**	**99.1%**	99.1%	**98.4%**	**98.8%**	98.5%	**89.5%**
	S-A	98.7%	**99.7%**	**99.1%**	99.0%	98.1%	98.4%	98.6%	89.2%
	RNN-A	98.6%	99.6%	98.8%	98.9%	97.6%	98.3%	98.4%	88.3%

Results on AOLP: As shown in Table 7.6, the proposed method outperform other methods on each dataset. Similar to results on CCPD, it has more advantages on tilted license plates, i. e., the RP subset.

Table 7.6 The performance of AOLP

Method		AC(681)	LE(757)	RP(611)
Li et al.		94.9%	94.2%	88.4%
Li et al.		95.3%	96.6%	83.7%
Wu et al.		96.6%	97.8%	91.0%
Zhang et al.		97.3%	98.3%	91.9%
Ke et al. (L)		97.7%	98.2%	87.9%
Ours small	PVAM	98.2%	99.1%	85.1%
	S-A	97.5%	97.9%	80.9%
	RNN-A	97.7%	95.3%	79.2%
Ours base	PVAM	98.5%	99.2%	95.5%
	S-A	97.5%	98.4%	91.0%
	RNN-A	98.1%	97.9%	94.1%

7.3.3 Ablation study

We chose the group 2 dataset, i. e., Rodosol-ALPR, for ablation experiments, as it contains abundant and balanced multi-type license plates. The effectiveness of proposed modules. To analyze the effectiveness of GMM and PCL, we conduct ablation experiments with different network structures. As shown in Table 7.7, "Baselin-ST-ST" and "Baseline-MT" respectively represent training baseline models using single type and multi type license plates. +GMM or +PCL represent models with added "GMM" or "PCL".

GMM can significantly improve performance, proving that proper global information is important for accurately perceiving the arrangement of characters to predict correct results. Notably, plugging in the GMM module only brings little cost. The number of parameters increases by about 100k/400k for models with the light/base encoder. Besides, PCL can further enhance performance, demonstrating that PCL can enhance the discriminability of visual features, especially for characters with similar glyphs. In addition, single-type LPR models cannot accurately recognize other license plate types, so an ensemble of them is required to achieve multi-type LPR, which consumes storage space. However, the proposed method is only a unified model and surpasses all single-type baselines, proving the superiority of the proposed method.

Table 7.7 The ablation study on the proposed modules

Model			Avg	Cars-br	Cars-me	Motor-br	Motor-me
Ours small	PVAM	Baseline-ST	92.7%	95.2%	98.0%	88.6%	88.9%
		Baseline-MT	88.5%	92.7%	95.3%	82.1%	84.0%
		+GMM	94.6%	95.5%	97.8%	91.7%	93.7%
		+PCL	95.3%	96.5%	98.6%	91.6%	94.4%
	S-A	Baseline-ST	91.7%	94.2%	97.6%	85.8%	89.2%
		Baseline-MT	92.2%	94.2%	97.5%	87.4%	89.7%
		+GMM	92.4%	94.4%	96.8%	87.5%	91.0%
		+PCL	93.9%	95.8%	98.2%	89.4%	92.2%
	RNN-A	Baseline-ST	88.3%	88.4%	93.8%	81.9%	89.0%
		Baseline-MT	58.2%	80.4%	85.5%	27.7%	39.2%
		+GMM	93.6%	95.5%	97.3%	88.9%	92.8%
		+PCL	93.2%	95.6%	97.7%	87.3%	92.1%
Ours base	PVAM	Baseline-ST	93.7%	94.9%	97.6%	90.8%	91.5%
		Baseline-MT	96.1%	97.6%	98.8%	92.7%	95.5%
		+GMM	96.5%	97.7%	98.8%	93.3%	96.3%
		+PCL	96.9%	98.1%	99.3%	94.0%	96.2%
	S-A	Baseline-ST	93.9%	94.8%	98.6%	90.6%	91.7%
		Baseline-MT	95.6%	97.3%	98.5%	91.7%	95.1%
		+GMM	96.0%	97.4%	98.6%	92.8%	95.4%
		+PCL	96.6%	98.2%	99.0%	94.0%	95.4%
	RNN-A	Baseline-ST	95.0%	96.1%	98.1%	92.4%	93.4%
		Baseline-MT	94.1%	96.4%	97.5%	89.4%	93.1%
		+GMM	96.1%	97.0%	98.4%	93.4%	95.4%
		+PCL	96.2%	97.4%	98.7%	93.4%	95.4%

The effectiveness of position-aware in PCL. We compared the proposed PCL with vanilla contrastive learning to verify the effect of position-aware strategy. Vanilla contrastive learning does not take the position of characters into consideration, and it directly pulls the character features of the same category closer and pushes the character features of different categories further. As shown in Table 7.8, the proposed method achieves better performance in almost all configurations. This indicates that distinguishing the position of characters is beneficial for multi-type LPR because there is a fixed template for each type of license plate.

Table 7.8 The effect of whether to consider the position of characters in contrast learning

Model		Position	Avg	Cars-br	Cars-me	Motor-br	Motor-me
Ours small	PVAM		94.5%	96.0%	98.4%	89.6%	94.2%
		√	95.3%	96.5%	98.6%	91.6%	94.4%
	S-A		93.5%	95.2%	97.2%	89.6%	92.2%
		√	93.9%	95.8%	98.2%	89.4%	92.2%
	RNN-A		91.3%	93.2%	95.5%	86.8%	89.8%
		√	93.2%	95.6%	97.7%	87.3%	92.1%
Ours base	PVAM		97.0%	98.0%	99.2%	94.6%	96.4%
		√	96.9%	98.1%	99.3%	94.0%	96.2%
	S-A		95.6%	97.3%	98.5%	91.7%	95.1%
		√	96.6%	98.2%	99.0%	94.0%	95.4%
	RNN-A		95.8%	97.1%	98.2%	93.0%	94.9%
		√	96.2%	97.4%	98.7%	93.4%	95.4%

Ablation study of λ. As shown in Equation (7.7), the total loss function has a hyperparameter λ to balance the two loss terms. We train the LPR models with the small encoder and different λ and test on the valid set of the group 2 dataset to find the suitable λ. As shown in Table 7.9, the best performance is achieved when $\lambda = 0.1$, so we use $\lambda = 0.1$ to train LPR models on other datasets.

Table 7.9 The ablation study of λ in loss

Method	λ	AVG	Cars-br	Cars-me	Motor-br	Motor-me
PVAM	0.01	95.2%	96.1%	99.2%	91.5%	94.0%
	0.1	95.3%	96.5%	98.6%	91.6%	94.4%
	1	95.2%	95.3%	99.1%	90.9%	95.4%
S-A	0.01	93.3%	93.9%	98.0%	87.4%	94.0%
	0.1	93.9%	95.8%	98.2%	89.4%	92.2%
	1	92.8%	94.7%	98.2%	88.5%	90.0%
RNN-A	0.01	90.8%	93.5%	95.3%	85.9%	88.6%
	0.1	93.2%	95.6%	97.7%	87.3%	92.1%
	1	93.4%	95.5%	97.2%	88.6%	92.4%

7.3.4 Attention visualization

The attention weights can indicate whether the decoder can accurately perceive the location of the characters, and we visualized the attention maps of the baseline model and our proposed model. Specifically, the attention weights obtained from the inference of LPR models are two-dimensional vectors. Thus, we directly normalize these vectors and convert them into pseudo-color images using open CV, and finally superimpose them with the input image. As shown in Fig. 7.6, our proposed method can better perceive the location of characters and achieve better results. The top row is the input license plate image. The 2nd-9th row is the attention visualization of the 1st-8th predicted character. The attention of the baseline may be misaligned with the corresponding character, causing wrong predictions. In comparison, our method utilizes GMM to introduce global features to the decoder, which enables the model to generate more precise attention, leading to more accurate predictions.

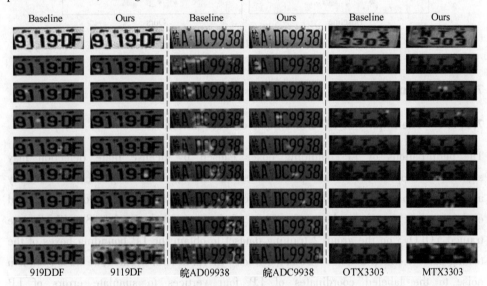

Fig. 7.6 The visualization of attention maps

7.3.5 T-sne visualization

To investigate the effect of the proposed PCL on characters with similar glyphs, the character features are visualized by utilizing the T-SNE algorithm. As shown in Fig. 7.7, each color stands for a character category. Compared to baseline, our proposed method has clearer classification boundaries, indicating it is easier to distinguish characters with similar glyphs. Characters with similar glyphs of the baseline model will be entangled in the feature space. Contrary to this, the proposed model has clearer classification boundaries, thus alleviating character misidentification. In addition, since PCL focuses not only on character classes but also on character positions, character features of the same category have different class centers that indicate the positions of characters.

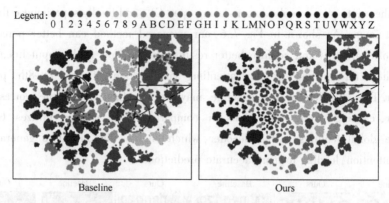

Fig. 7.7 The T-SNE visualization of character features

7.3.6 Discussion

7.3.6.1 The impact of LP Bbox noise on recognition

The perturbations of LP bounding boxes caused by license plate detection may significantly impact recognition performance. Therefore, we investigated this issue by adding noise to the GT bounding boxes(Bboxes) or using detection bounding boxes and evaluating how it affects the performance of the recognition models. Specifically, we trained YOLOv3 on the group 2 training dataset to obtain the LP bounding boxes of the testing dataset, which denotes "Detected Bbox" in Table 7.8. Besides, we added random noise to the labeled coordinates of LP four vertices to simulate errors of LP detection. The noise intensity is classified according to the ratio of the absolute value of the changes in the horizontal/vertical coordinates to the LP width/height. For example, the "0-1%" in Fig. 7.8 represents the horizontal/vertical coordinates of LP vertexes randomly increasing or decreasing a value, which is 0-1% of the LP width/height. As shown in Fig. 7.8, the proposed method is more robust to the Bboxes noise than the baseline models. In addition, it can be seen that the recognition performance using detected Bboxes is better than the performance on the annotated LP Bboxes with manual noise, which indicates that the detection model is able to locate LPs precisely and has less impact on the recognition performance.

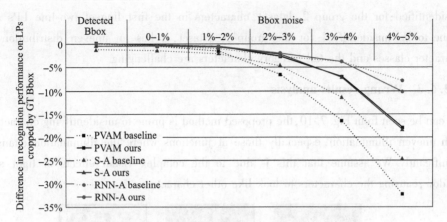

Fig. 7. 8　The impact of LP Bbox noise on recognition

7.3.6.2　Challenging character analysis

As shown in Fig. 7.9, we visualized the characters in each dataset by their character error rate to find out which character is more challenging. The recognition model used is the model based on the base encoder and PVAM decoder, which achieves the best performance among the proposed models. Based on the analysis of wrong samples, we believed that for the group 1 dataset, characters with complex glyphs are prone to be

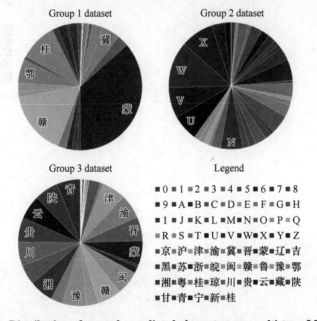

Fig. 7. 9　Distribution of wrongly predicted characters on multi-type LP dataset

misidentified; for the group 2 dataset, characters in the first line of two-line LPs are prone to be misidentified; for the group 3 dataset, it has an uneven distribution of character classes and the tail category characters are challenging.

7.3.6.3 Wrong results analysis

As can be seen from Fig. 7.10, the proposed method is prone to misidentifying characters with uneven illumination, especially those at junctions where the illumination changes significantly. We assume that this is due to the complex visual effects of light and shadow, causing the characters to look like other characters.

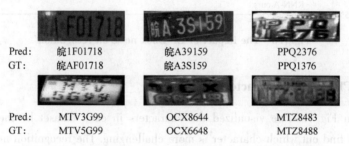

Pred:	皖1F01718	皖A39159	PPQ2376
GT:	皖AF01718	皖A3S159	PPQ1376

Pred:	MTV3G99	OCX8644	MTZ8483
GT:	MTV5G99	OCX6648	MTZ8488

Fig. 7.10 Examples of wrongly recognized license plates

Please see colorful images

Chapter 8 Towards Low-resource License Plate Recognition via Feature Shuffling

Manual annotation is costly and limits the availability of sufficient annotated license plates for training recognition models. Small-scale license plate datasets (i. e. , low-resource) often exhibit a long-tailed distribution in character classes at some character positions, primarily due to their limited variation in character permutations. Previous methods tend to prioritize head classes with high occurrence probability when applied to small-scale datasets. To solve this problem, we propose feature shuffling to balance the occurrence distribution across various character classes, thereby improving the recognition of tail classes with low occurrence probability. Moreover, we introduce global perception to holistically understand the overall character layout for effective feature shuffling. Extensive experiments on the small-scale UFPR and SSIG datasets demonstrate that our method achieves state-of-the-art results, with an average improvement of 43.70% over the baseline. Experiments on RodoSol and CCPD prove our method achieves state-of-the-art performance on large-scale datasets, verifying its generality.

8.1 Problem formulation

License plate recognition is vital for intelligent transportation systems, widely used in parking management and electronic toll collection. Despite excelling with large-scale datasets, many sequence recognition methods struggle when applied to small-scale datasets. In essence, they face difficulties in low-resource license plate recognition.

Given an ample number of training samples, the character classes at each character position tend to exhibit a uniform distribution. In other words, the occurrence probability of all character classes at each position is approximately equal. However, observations indicate that training samples within small-scale datasets are constrained and characterized by limited variation in character permutations. The occurrence probability of character classes at some positions exhibits a long-tailed distribution. That is, only the head classes have a higher occurrence probability at certain positions, and the tail classes have a small occurrence probability. As shown in Fig. 8.1, we take the second character position in the SSIG training set as an example. The small number of samples in SSIG and the limited variety of character permutations lead to a long-tailed distribution of character classes at this character position. Evidently, class "P" is predominant, whereas classes "D", "G", and "R" show no occurrences. In the SSIG testing set, we observed that the model tends to recognize the character at this position as the head class "P", which indicates the network prioritizes head classes. A similar phenomenon also occurs for other character positions, resulting in incorrect recognition results.

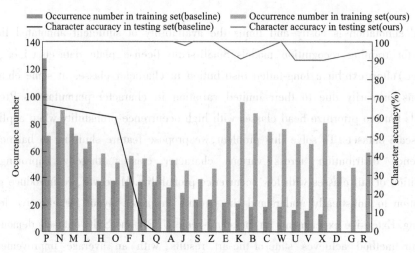

Fig. 8. 1 The occurrence distribution histograms of character classes and recognition accuracy curves of characters

To address this issue, character detection methods can directly detect and recognize each character independently, disregarding its position within the license plate string. These methods essentially bypass the challenge posed by the long-tailed distribution of character classes. However, character detection methods require character-level annotations, wasting manpower and resources. Moreover, image generation methods can enhance low-resource license plate recognition by generating a substantial number of diverse synthetic license plates. The generated license plates can enhance the training datasets by offering additional training samples and ensuring a more uniform distribution of character classes at each position. However, image generation struggles to produce realistic images effectively when using small-scale datasets. Furthermore, image generation necessitates an additional generator network, thereby escalating computational resources and training time. To enhance low-resource license plate recognition, we propose to balance the occurrence distribution across various character classes through feature shuffling. As illustrated in Fig. 8. 1, our approach effectively mitigates the long-tailed distribution of character classes, significantly improving the recognition of tail classes. Specifically, we employ the reading order as the guidance information to spatially attend to characters while implementing feature shuffling through the shuffling of the reading order. We can obtain a balanced occurrence distribution of various character classes at each position through feature shuffling. To ensure effective feature shuffling, we introduce global perception to holistically understand the overall character layout within the license plate. Our proposed method relies solely on string-level annotations for sequence recognition, eliminating the need for character-level annotations as required by character detection methods. Furthermore, our approach achieves a similar outcome to image generation methods in terms of providing additional training samples, without the necessity of an extra generator network. Extensive experiments demonstrate that our method achieves state-of-the-art performance, with improvements of 38.39% and 49.01% on the small-scale UFPR and SSIG datasets compared to the baseline model, respectively. Notably, our approach demonstrates exceptional performance on large-scale datasets, achieving state-of-the-art performance on RodoSol, CCPDv1, and CCPDv2.

8.2 Methodology

As shown in Fig. 8.2, our method comprises three key components, i.e., the Global Perception Encoder(GPE), the Feature Shuffling Module(FSM), and the Decoder.

GPE integrates local perception and global perception to extract comprehensive global visual features. The local perception network utilizes convolutions to extract specific local visual features. The global perception network further refines these features, facilitating a holistic understanding of the overall character layout and enhancing subsequent feature shuffling. More specifically, the local perception network comprises nine convolutional layers. Following the third and sixth convolutions, a 2 × 2 max-pooling layer is applied, resulting in a down-sampling rate of $r = 4$. Each convolutional layer employs a 3 × 3 kernel with a stride of 1, followed by Batch Normalization and ReLU activation. The input image is initially converted to grayscale, maintaining a fixed size ($H \times W \times C$) of 32× 96 × 1. The local perception network generates a feature map of 8 × 24 × 128, representing the local visual features. The global perception network uses these local visual features as query, key, and value for feature interaction, enabling global information mining. Specifically, we adopt the vanilla transformer encoder, which integrates two 4-head self-attention layers, a feed-forward layer with a dimension of $d_{ff} = 512$, and an output layer with a dimension of $d_{model} = 128$. After performing global perception, we obtain the global visual features $F_G \in R^{(\frac{H}{r} \times \frac{W}{r})}$.

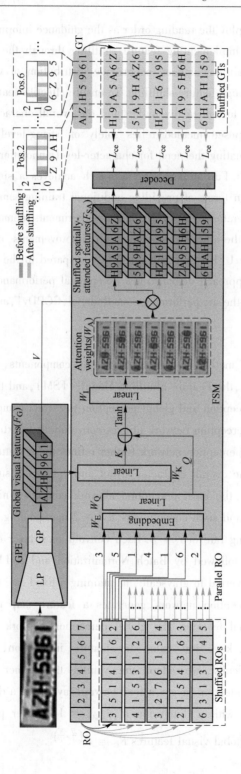

Fig. 8.2 Overall architecture

Based on the global visual features from GPE, we employ the reading order as the guidance information to enhance character recognition by spatially attending to characters. Additionally, to balance the occurrence distribution across various character classes at each position, we shuffle the original reading order into multiple new reading orders. Using these shuffled reading orders facilitates feature shuffling, thereby generating new character permutations. As illustrated in Fig. 8.2, the original permutation is "AZH5961" with an initial reading order spanning from 1 to 7. After undergoing five random shufflings, we generate alternative reading orders like "3514162" with a corresponding permutation "H9A5A6Z". This way, we alleviate the issue of long-tailed distribution of character classes, thus improving recognition of tail classes.

Spatial Character Attention: In essence, spatial character attention is executed by establishing a mapping between the query (Q) and the key (K) to derive attention weights. Subsequently, the value (V) is weighted and aggregated using these attention weights, resulting in spatially-attended features that enhance character recognition. In this work, we utilize the shuffled reading order as the query, while the key-value set is derived from the global visual features F_G extracted by GPE. Specifically, the query originates from the parallel reading order $RO \in R^T$, where $T=8$ denotes the maximum length of license plate strings. To ensure uniformity in the length of all plates, we append multiple blank characters to plates with string lengths shorter than T. Initially, parallel embedding encoding W_E is applied to RO, resulting in $F_E \in R^{T \times d_h}$, where $d_h = 128$ represents the hidden dimension. Following this, F_E is projected to the query embedding $Q \in R^{(\frac{H}{r} \times \frac{W}{r})}$ via a linear layer W_Q. Similarly, $K \in R^{(\frac{H}{r} \times \frac{W}{r})}$ are calculated based on Q and K. The value, denoted as V, remains consistent with F_G. Finally, the spatially-attended features $F_{SA} \in R^{T \times d_{model}}$ are generated utilizing the attention weights and V. The detailed computational process is outlined from Equation(8.1) to Equation(8.3).

$$Q = (W_E(RO)^T \times W_Q)^T, \quad K = F_G \times W_K, \quad V = F_G \qquad (8.1)$$

$$W_A = \text{Softmax}((\text{Tanh}(Q + K) \times W_I)^T) \qquad (8.2)$$

$$F_{SA} = W_A \times V \qquad (8.3)$$

where, $W_K \in R^{d_{model} \times d_h}$, $W_Q \in R^{T \times (\frac{H}{r} \times \frac{W}{r})}$, $W_E \in R^{T \times d_h}$, and $W_I \in R^{d_h \times T}$ are trainable parameters. The superscript T represents matrix transpose.

Reading Order Shuffling: As depicted in Fig. 8.2, we randomly shuffle the reading order five times for each license plate in every training epoch. This strategy contributes to a more balanced occurrence distribution of various character classes at each character position, leading to a significant enhancement in the recognition of tail classes, as illustrated in Fig. 8.1. Taking the SSIG training set as an example (800 samples), the reading order shuffling strategy ideally generates approximately 3.2 million (800×5× 800) training samples after training for 800 epochs.

The most straightforward shuffling strategy involves Random Selection (RS) of a reading order for each position. However, RS is more suitable for large-scale datasets with uniformly distributed character classes, as it does not consider the occurrence distribution in the training sets. In the case of small-scale datasets, certain characters, particularly the tail classes across all character positions, have fewer occurrences and thus a smaller probability of being shuffled. Consequently, even after reading order shuffling, these tail classes maintain a small occurrence probability. To tackle this issue, we propose Considering the Probability(CP) of character occurrence in the training sets across all character positions. The primary goal is to amplify the occurrence probability of tail classes and establish a more uniform occurrence probability for all character classes. Specifically, for SSIG, which comprises $NC = 37$ classes including 26 English letters, 10 Arabic digits, and a blank character, our aim is to achieve an occurrence probability of $1/NC$ for all character classes at each position. CP involves diminishing the occurrence probability of head classes while elevating the occurrence probability of tail classes. Furthermore, during training, a batch of license plates is simultaneously used for training. Utilizing the same shuffled reading order for the entire batch might restrict the diversity of training samples. Hence, we propose Independently Shuffling (IS) each license plate within the same batch to enhance the variety of character permutations. IS aims to generate a more uniform occurrence distribution across the dataset.

Following the shuffled reading order, we can attend to the corresponding characters of the new permutation within the global visual features F_G. Consequently, we can generate shuffled spatially-attended features F_{SA} along with their associated ground truth, aligned with the shuffled reading order. The feature shuffling process can enhance the recognition of tail classes due to the more balanced occurrence distribution of character classes at each position.

As license plates lack semantic relations between characters, we simply use a fully connected layer as the decoder to predict all characters in parallel. As depicted in Equation(8.4), we employ cross-entropy loss for training.

$$L_{ce} = -\frac{1}{T}\sum_{t=1}^{T} y_t \lg(p_t) \tag{8.4}$$

where, p_t and y_t represent the predicted probability and ground truth at the decoding step t, respectively.

8.3 Experiments

We conduct experiments on four public datasets: UFPR, SSIG-SegPlate, RodoSol, and CCPD. UFPR and SSIG are small-scale datasets comprising 4500 and 2000 images, respectively. RodoSol and CCPD are large-scale datasets, with RodoSol containing 20k images. CCPD has two versions: CCPDv1 with 250k images and CCPDv2 with 300k images.

GP and ROS: In Table 8.1, the Global Perception(GP) network in our proposed GPE demonstrates a 3.61% improvement on UFPR and 15.05% on SSIG compared to the baseline, respectively.

Table 8.1 Ablation study on Global Perception(GP) and Reading Order Shuffling(ROS)

GP	ROS	UFPR	SSIG	Params.	FPS
		7.33%	8.58%	0.506M	209.17
√		10.94%	23.63%	0.902M	157.39
	√	6.22%	16.29%	0.506M	209.17
√	√	**45.72%**	**57.59%**	0.902M	157.39

The Reading Order Shuffling(ROS) process performed in FSM provides an additional boost of 34.78% on UFPR and 33.96% on SSIG, surpassing the baseline by 38.39% and 49.01%, respectively, without increasing parameters or FPS. Notably, applying only ROS without GP results in a slight performance decline on UFPR compared to the baseline, highlighting the enhanced effectiveness of ROS when used in conjunction with GP.

Reading Order Shuffling Strategies: In Table 8.2, we conduct ablation experiments on the reading order shuffling strategies proposed. The improvements from Random Shuffling(RS) highlight the effectiveness of reading order shuffling in achieving a more balanced occurrence distribution for character classes at each position. Considering the Probability(CP) of character occurrence in the training sets further enhances recognition performance by increasing the occurrence probability of tail classes. Additionally, Independently Shuffling(IS) each sample in the same training batch improves recognition performance by introducing more diverse character permutations, thereby balancing the occurrence distribution across various character classes.

Table 8.2 Ablation study on reading order shuffling strategies

RS	IS	CP	UFPR	SSIG
			10.94%	23.63%
√			26.28%	40.30%
√	√		43.61%	43.16%
√	√	√	**45.72%**	**57.59%**

In Fig. 8.3, our method effectively balances the occurrence distribution of character classes in the second character position of the UFPR training set. This balanced distribution enhances the recognition of tail classes at this position, akin to the effects observed in Fig. 8.1 for SSIG. For the occurrence distribution histograms of character classes and character accuracy curves on RodoSol and CCPD.

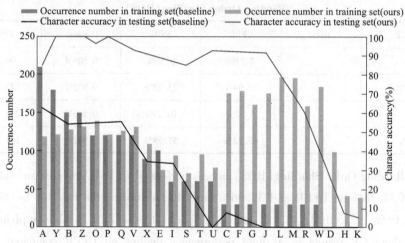

Fig. 8.3 The occurrence distribution histograms of character classes and character accuracy curves on UFPR

UFPR and SSIG: Table 8.3 demonstrates that our method achieves the best recognition performance with the fewest parameters on UFPR and SSIG.

Table 8.3 Comparative experiments on UFPR and SSIG

Method	UFPR		SSIG-Segplate		Params
	Real	Real+Syn	Real	Real+Syn	
Sequence recognition					
RNN-Attention	0.00%	67.39%	2.61%	85.45%	1.265M
PVAM	0.00%	66.11%	2.61%	83.96%	5.345M
HC	3.33%	61.67%	2.61%	79.48%	6.812M

Continued Table 8.3

Method	UFPR		SSIG-Segplate		Params
	Real	Real+Syn	Real	Real+Syn	
Sequence recognition					
DAN	13.67%	70.94%	13.43%	87.81%	2.629M
LPRNet	12.39%	47.49%	18.10%	75.39%	1.916M
Gonçalves et al.	—	55.60%	—	85.60%	—
Character detection					
SSD	5.44%	—	15.42%	—	28.424M
YOLO	38.17%	—	56.47%	—	61.712M
Montazzolli et al.	—	—	—	63.10%	—
Laroca et al.	—	64.90%	—	85.50%	—
Silva et al.	—	72.20%	—	88.60%	—
Sighthound	—	47.40%	—	73.10%	—
OpenALPR	—	57.90%	—	87.40%	—
Ours	45.72%	76.72%	57.59%	90.55%	0.902M

Sequence recognition methods encounter significant challenges on these small-scale datasets, particularly RNN-Attention and PVAM. These methods tend to heavily overfit the limited character permutations in the training set, where a long-tailed distribution of character classes exists at each position. Consequently, the discrepancy in the distribution of character classes between the training and testing sets leads to lower recognition performance. Although character detection methods can enhance recognition by bypassing the long-tailed distribution of character classes, they often require a considerable number of parameters. Additionally, our method significantly improves recognition performance when using synthetic license plates by introducing more diverse character permutations and ensuring a more uniform distribution of character classes at each position. This enhancement allows us to achieve state-of-the-art recognition performance. For the generation process of synthetic license plates.

CCPDv1 and CCPDv2: As depicted in Table 8.4, our method outperforms the previous state-of-the-art method by 7.3% across all subsets of CCPDv2 and achieves the best performance on all subsets, demonstrating its generality on large-scale datasets. Furthermore, we demonstrate the effectiveness of GP and ROS on large-scale datasets. GP aids in holistically understanding the character layout, while ROS establishes a balanced distribution of character classes. Notably, ROS maximizes its efficacy only when combined with GP.

Table 8.4　Comparative experiments on CCPDv2

Method	Avg	DB	Blur	FN	Rot.	Tilt	Cha.
Xu et al. (2018)	43.4%	34.5%	25.8%	45.2%	52.8%	52.0%	44.6%
Zherzdev et al. (2018)	69.0%	63.0%	59.2%	71.9%	73.1%	62.8%	74.6%
Luo et al. (2019)	64.2%	55.5%	57.6%	66.1%	70.1%	55.9%	69.3%
Wang et al. (2020)	46.8%	43.9%	39.9%	52.6%	50.4%	35.5%	53.4%
Wang et al. (2022)	42.1%	38.2%	36.1%	51.0%	36.5%	36.6%	46.2%
Li et al. (2022)	67.8%	67.3%	75.3%	67.0%	63.3%	47.5%	78.4%
Chen et al. (2023)	73.5%	63.9%	63.0%	75.2%	75.7%	73.4%	74.6%
Ke et al. (2023)	74.8%	68.3%	69.2%	77.5%	77.9%	72.1%	78.3%
Ours(baseline)	78.4%	73.1%	72.1%	80.7%	82.8%	77.3%	80.9%
Ours(with GP)	79.1%	74.0%	71.5%	81.7%	84.4%	79.5%	80.9%
Ours(with ROS)	79.7%	75.9%	72.4%	81.9%	84.6%	80.0%	81.5%
Ours(with GP and ROS)	82.1%	78.3%	75.5%	84.0%	87.3%	82.7%	83.3%

Table 8.5 demonstrates that our method achieves state-of-the-art performance across all subsets of CCPDv1 and achieves the best performance on all subsets, verifying its generality on large-scale datasets as CCPDv2.

Table 8.5　Comparative experiments on CCPDv1

Method	Avg	Base	DB	FN	Rot.	Tilt	Wea.	Cha.
Zhang et al. (2016)	93.0%	99.1%	96.3%	97.3%	95.1%	96.4%	97.1%	83.2%
Joseph et al. (2017)	93.7%	98.1%	96.0%	88.2%	84.5%	88.5%	87.0%	80.5%
Zherzdev et al. (2018)	93.0%	97.8%	92.2%	91.9%	79.4%	85.8%	92.0%	69.8%
Xu et al. (2018)	95.5%	98.5%	96.9%	94.3%	90.8%	92.5%	87.9%	85.1%
Li et al. (2019)	94.4%	97.8%	94.8%	94.5%	87.9%	92.1%	86.8%	81.2%
Luo et al. (2019)	98.3%	99.5%	98.1%	98.6%	98.1%	98.6%	97.6%	86.5%
Wang et al. (2020)	96.6%	98.9%	96.1%	96.4%	91.9%	93.7%	95.4%	83.1%
Zhang et al. (2021)	98.5%	99.6%	98.8%	98.8%	96.4%	97.6%	98.5%	88.9%
Liu et al. (2021)	98.7%	99.7%	99.0%	99.2%	97.7%	98.3%	98.8%	89.2%
Zou et al. (2022)	97.5%	99.2%	98.1%	98.5%	90.3%	95.2%	97.8%	86.2%
Chen et al. (2023)	98.8%	99.7%	99.1%	98.8%	98.4%	98.5%	98.8%	90.0%
Ours(baseline)	99.3%	99.9%	99.5%	99.7%	99.2%	99.3%	99.3%	92.6%

Continued Table 8.5

Method	Avg	Base	DB	FN	Rot.	Tilt	Wea.	Cha.
Ours(with GP)	99.4%	99.9%	99.6%	99.7%	99.4%	99.5%	99.3%	93.3%
Ours(with ROS)	99.2%	99.8%	99.5%	99.6%	99.0%	99.2%	99.1%	92.1%
Ours(with GP and ROS)	99.5%	99.9%	99.8%	99.8%	99.5%	99.6%	99.5%	94.3%

Please see colorful images

Chapter 9 End-to-End Multi-line License Plate Recognition with Cascaded Perception

Due to the irregular layout, multi-line license plates are challenging to recognize, and previous methods cannot recognize them effectively and efficiently. To solve this problem, we propose an end-to-end multi-line license plate recognition network, which cascades global type perception and parallel character perception to enhance recognition performance and inference speed. Specifically, we first utilize self-information mining to extract global features to perceive plate type and character layout, improving recognition performance. Then, we use the reading order to attend plate characters parallelly, strengthening inference speed. Finally, we propose extracting recognition features from shallow layers of the backbone to construct an end-to-end detection and recognition network. This way, it can reduce error accumulation and retain more plate information, such as character stroke and layout, to enhance recognition. Experiments on three datasets prove our method can achieve state-of-the-art recognition performance, and cross-dataset experiments on two datasets verify the generality of our method. Moreover, our method can achieve a breakneck inference speed of 104 FPS with a small backbone while outperforming most comparative methods in recognition.

9.1 Problem formulation

License Plate Recognition (LPR) is essential to intelligent transportation systems like traffic supervision and vehicle management. However, as shown in Fig. 9.1, multi-line license plates are challenging to recognize because they are irregular in character layouts compared with single-line license plates.

Fig. 9.1 Double-line and single-line license plates

As shown in Fig. 9.2, previous multi-line LPR methods can be roughly categorized into three types, (a) character segmentation, (b) line segmentation, and (c) (d) segmentation-free. Firstly, like Fig. 9.2 (a), Silva et al. propose to detect each plate character independently. However, these methods require labor-intensive annotations and suffer character detection errors. Secondly, like Fig. 9.2(b), Cao et al. propose to bisect the double-line LP horizontally and then splice it into a single-line LP for recognition. However, these methods cannot handle tilted license plates due to incorrect bisection. Thirdly, researchers propose to spatially attend plate characters without segmentation, including CNN-based and RNN-based. However, like Fig. 9.2 (c), the CNN-based methods can only extract local features and lack the overall perception of plate type and character layout, which causes attention errors and affects recognition performance. Like Fig. 9.2(d), the RNN-based methods can pre-classify plate type as the prior knowledge, but they are time-consuming due to encoding features step by step. To solve the above problems, like Fig. 9.2(e), we propose to cascade global type perception and parallel character perception to improve multi-line LPR. Firstly, we use self-information mining to extract global features to pre-perceive plate type and character layout, which can improve multi-line LPR by enhancing the subsequent character attention. Then, we use the reading order to spatially attend plate characters, which can parallelly encode features to enhance the inference speed. Finally, we can accurately recognize multi-line license plates with globally and spatially enhanced features.

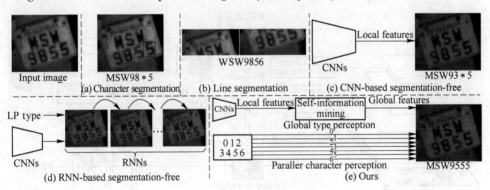

Fig. 9.2　Our proposed method

Moreover, previous methods utilize separate networks for LPR, where the detected license plates are extracted from the original image for recognition. However, these methods could cause error accumulation because the sub-sequent recognition will inevitably fail if the previous detection fails. To reduce error accumulation, researchers propose end-to-end methods to jointly optimize detection and recognition, where recognition features are extracted from deep layers of the backbone. However, deep-layer features are not conducive to LPR due to small-sized features after multiple down-sampling operations. To solve this problem, we propose to extract recognition features from shallow layers of the backbone because shallow-layer features retain more plate information for LPR, such as character stroke and layout.

Extensive experiments on RodoSol, UFPR, and CCPD prove our method can achieve state-of-the-art recognition performance, especially for multi-line license plates. We also conduct cross-dataset experiments to verify the generality of our proposed method, i.e., training on CCPD and testing on PKUData and CLPD. Moreover, our method can achieve the fastest inference speed of 104 FPS with a small backbone while outperforming most comparative methods in recognition performance.

9.2 Methodology

Fig. 9.3 illustrates the proposed network, including the backbone, detection head, recognition feature extraction, and recognition. Firstly, the backbone can extract shared features for detection and recognition. Then, the detection head can accurately detect the license plate using multiple deep-layer features. Finally, the license plate can be accurately recognized with cascaded perception, where the recognition features are extracted from the first convolutional layer of the backbone to construct an end-to-end network.

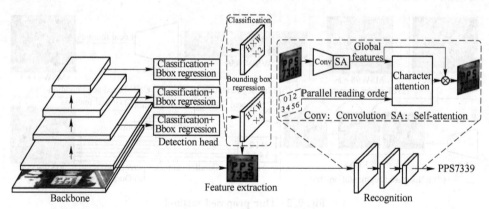

Fig. 9.3 Overall architecture

9.2.1 Backbone

We use the large VGGNet or small MobileNetV1 as the backbone to evaluate the generality, which can extract shared features for detection and recognition. For VGGNet, we use the same network as the vanilla SSD for a fair comparison. For MobileNetV1, we adopt the network implemented by the third party. The input size of the backbone is set to 512×512. If not specified, VGGNet is used by default.

9.2.2 License plate detection

We use the prestigious SSD as the detector and replace the bounding box regression loss from Smooth L_1 loss to CIoU loss. This way, it can enhance detection because CIoU loss considers the overlap area, central point distance, and aspect ratio for better and faster regression. The training loss comprises the classification loss L_{cls} and bounding box regression loss $L_{CIoU}(B, B_{gt})$.

$$L_{det} = \frac{1}{N}(L_{cls} + \alpha L_{CIoU}(B, B_{gt})) \quad (9.1)$$

where, L_{cls} is implemented by Softmax loss. N denotes the number of matched anchor boxes. α represents the loss balance term and is set to 5 by default. B and B_{gt} denote the predicted box and ground-truth box, respectively.

9.2.3 Recognition feature extraction

Generally, deep-layer features have strong semantics and a large receptive field, which are conducive to LPD. However, deep-layer features are not conducive to LPR because they undergo multiple down-sampling operations and lose much information, such as character stroke and layout. On the contrary, shallow-layer features have weak semantics and a large resolution, which retain the information of character stroke and layout to enhance recognition. In this work, we propose to extract recognition features from the first convolutional layer of the backbone and construct the network end-to-end. This way, it can reduce error accumulation and achieve the best recognition performance.

9.2.4 License plate recognition

After recognition feature extraction, we can get plate features $X_{96\times32}$ with the size of 96× 32, which is the input of the recognition network. As shown in Fig. 9.3, the proposed recognizer consists of local feature extraction, global type perception (GTP), parallel character perception (PCP), and character prediction. GTP and PCP are implemented by self-attention and spatial attention, respectively.

Local Feature Extraction. The extracted recognition features $X_{96\times32}$ can generate local features $X_{24\times8}$ through multi-layer convolutions. As shown in Equation (9.2), there are eighteen convolutional layers, including six convolutions of 64, 128, and 256 channels, respectively. All convolutional layers consist of Convolution, BatchNorm, and ReLU. There are two down-sampling operations after the sixth and twelfth layers, so the size of local features is 24×8.

$$X_{24\times8} = [\text{ReLU}(\text{BN}(\text{Conv}(X_{96\times32})))]^{\times18} \quad (9.2)$$

Global Type Perception. However, the local features $X_{24\times8}$ cannot perceive plate type and character layout, causing character attention errors. As shown in Fig. 9.4, we use GTP to extract global features $Y_{24\times8}$, enhancing the subsequent character attention. The global features are generated via self-information mining, and the key, query, and value are all generated by local features. Specifically, we adopt the vanilla transformer unit and stack two transformer units with attention heads $h=4$, feed-forward dimension $d_{\text{ff}} = 512$, and output dimension $d_{\text{model}} = 256$. W_K, W_Q, W_V, and W denote learnable parameters for the key, query, value, and position-wise feed-forward network, respectively.

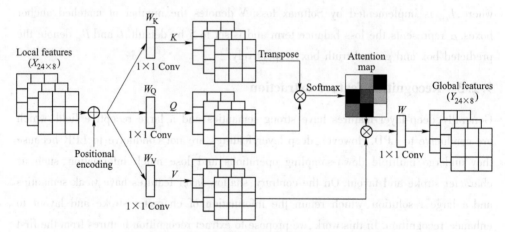

Fig. 9.4 The architecture of global type perception

Parallel Character Perception. As shown in Fig. 9.5, we use the reading order to parallelly attend plate characters of different layouts, thus improving the inference speed. This way, the global features $Y_{24\times8}$ can generate spatially-attended features $Z_{\text{max}T} \times d_{\text{model}}$, where $\text{max}T = 8$ denotes the maximum character length of the license plate. Specifically, the key-value set is generated from global features $Y_{24\times8}$, and V is the same as $Y_{24\times8}$, without any extra operation. The query is the parallel reading order, which is first transformed into one-hot vectors and then encoded by parallel embedding with the hidden dimension $dh = 256$. W_K, W_Q, W_E, and W represent learnable parameters for the key, query, embedding function, and output function, respectively. The embedding function can be regarded as a fully-connected layer without bias.

Character Prediction. Finally, we adopt a fully-connected layer to map the spatially-attended features $Z_{\text{max}T} \times d_{\text{model}}$ to the class probabilities $P_{\text{max}T \times C}$. As shown in Equation(9.3), we use the cross-entropy loss as the recognition loss to maximize the

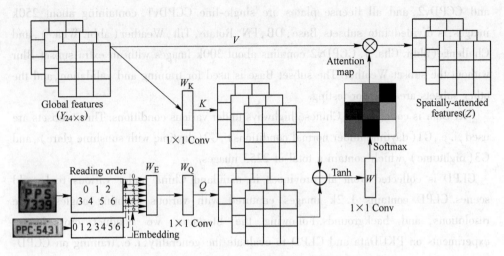

Fig. 9.5 The architecture of parallel character perception

prediction confidence of each character. Notably, we add blank characters at the end of the license plate for parallel training, which ensures that all license plates are aligned to the same character length of max$T$.

$$L_{rec} = -\sum_{t=1}^{\max T}\sum_{c=1}^{C} y_{tc}\lg(p_{tc}) \qquad (9.3)$$

where, C denotes the number of character categories. y and p represent the ground-truth label and predicted class probability, respectively.

The whole network can be trained in an end-to-end manner.

$$L = L_{det} + L_{resc} \qquad (9.4)$$

9.3 Experiments

9.3.1 Datasets and evaluation metrics

RodoSol is collected at toll booths on Brazilian highways, containing 20k images. RodoSol contains license plates of two layouts: single-line for cars and double-line for motorcycles. The number of single-line and double-line license plates is the same. We follow the official split, i. e., 8k, 4k, and 8k images for training, validation, and testing, respectively. UFPR is collected on Brazilian roads with driving recorders, containing 4.5k images. Like RodoSol, UFPR contains single-line and double-line license plates, where the number of single-line is three times that of double-line. We randomly select 40%, 20%, and 40% images for training, validation, and testing, respectively. CCPD is collected in roadside parking scenes of Chinese cities, including two versions: CCPDv1 and CCPDv2, and all license plates are single-line. CCPDv1, containing about 250k images, is divided into subsets Base, DB, FN, Rotate, Tilt, Weather(abbr. Wea.), and Challenge(abbr. Cha.). CCPDv2 contains about 300k images with an extra subset Blur without the subset Weather. The subset Base is used for training and validation, and the other subsets are used for testing.

PKUData is collected on Chinese highways under various conditions. Three subsets are used, i. e., G1(daytime under normal conditions), G2(daytime with sunshine glare), and G3(nighttime), which contain a total of 2253 images.

CLPD is collected from all provinces in mainland China under various real-world scenes. CLPD contains 1.2k images captured with various shooting angles, image resolutions, and backgrounds. Following the literature, we conduct cross-dataset experiments on PKUData and CLPD to evaluate the generality, i. e., training on CCPD-Base and testing on PKUData and CLPD.

We use text recognition rate to evaluate the recognition accuracy; that is, all characters must be recognized correctly. The recognition accuracy for single-line and double-line license plates is denoted as Single and Double, respectively. For RodoSol and UFPR, we also calculate the average accuracy(abbr. Avg) of single-line and double-line license plates to evaluate the overall performance. For CCPD, Avg denotes the average recognition accuracy of all the subsets. Moreover, we use frames per second (FPS) to evaluate the inference speed. Notably, there is only one license plate in each image of all the above datasets.

9.3.2 Ablation study

As shown in Table 9.1, we conduct ablation experiments of different network components on RodoSol, UFPR, and CCPD, including DE for detection, GTP and PCP for recognition, and E2E for the end-to-end network.

Table 9.1 Ablation study of different network components

Method	E2E?	RodoSol			UFPR			CCPDv2	CCPDv1	FPS
		Avg	Single	Double	Avg	Single	Double	Single	Single	
Ours	√	96.55%	98.08%	95.03%	99.67%	99.58%	100.00%	73.48%	98.8%	37
w/o DE	√	96.28%	97.60%	94.95%	98.22%	98.75%	96.15%	72.12%	98.7%	37
w/o GTP	√	93.94%	96.38%	91.50%	98.94%	99.16%	98.08%	70.03%	98.6%	38
w/o PCP	√	96.40%	97.85%	94.95%	99.00%	99.23%	98.08%	72.59%	98.7%	33
w/o E2E		93.94%	97.63%	90.25%	98.28%	98.47%	97.53%	67.32%	98.5%	37

E2E?: Is it an end-to-end license plate recognition network?

9.3 Experiments

DE (Detection Enhancement) means using CIoU loss for bounding box regression instead of Smooth L_1 loss in the vanilla SSD. CIoU loss regresses the box as a whole unit and considers the overlap area, central point distance, and aspect ratio to improve detection, thus enhancing recognition.

GTP (Global Type Perception) means the global perception of plate type and character layout. Although removing GTP can improve the inference speed by 1 FPS, it will significantly reduce the recognition accuracy, especially for the double-line license plate. For RodoSol and UFPR, the recognition accuracy of single-line license plates decreases by 1.70% and 0.42%, while that of double-line license plates decreases by 3.53% and 1.92%, respectively. As shown in Fig. 9.6, GTP can improve multi-line LPR by enhancing character attention.

Fig. 9.6 Character attention with or without GTP

PCP (Parallel Character Perception) means parallelly attending plate characters using the reading order. Without PCP, it will degrade to the RNN-based segmentation-free model, which sequentially encodes character features and is time-consuming. With PCP, our method can run 4 FPS faster due to encoding parallelly. Moreover, encoding the relationship between characters through RNNs will introduce noises because of no or minor semantics between plate characters. Our method can decouple the semantic relationship between characters by using reading orders independently, thus improving recognition.

E2E (End-to-End) means the proposed end-to-end network, i. e., extracting recognition features from the first convolutional layer of the backbone. Without E2E, it will consist of a separate detector and recognizer, which extracts the detected license plate from the original image. This way, it will significantly reduce recognition accuracy because of error accumulation, especially for double-line license plate. For RodoSol and UFPR, the recognition accuracy of single-line license plates decreases by 0.45% and 1.11%, while that of double-line license plates decreases by 4.78% and 2.47%, respectively.

As shown in Table 9.2, we study the effect of different recognition inputs, i.e., extracting the detected LP from the original image or convolutional layer. If from the original image, it will cause error accumulation to affect recognition; if from the convolutional layer, the network can be optimized jointly to improve recognition. Moreover, shallow convolutional layers are conducive to LPR because they are large-sized and retain more essential features, such as character stroke and layout. Deep-layer features are small-sized after multiple down-sampling operations, which are not conducive to LPR because of losing much information. In this work, we choose the first convolutional layer of the backbone as the recognition input, achieving the best recognition performance for all datasets.

Table 9.2   Ablation study of different recognition inputs

Input	E2E?	Size	RodoSol			UFPR			CCPDv2	CCPDv1
			Avg	Single	Double	Avg	Single	Double	Single	Single
Image		—	93.94%	97.63%	90.25%	98.28%	98.47%	97.53%	67.32%	98.5%
1st Conv	√	512	96.55%	98.08%	95.03%	99.67%	99.58%	100.00%	73.48%	98.8%
4th Conv	√	256	94.36%	96.05%	92.68%	99.28%	99.09%	100.00%	73.31%	98.8%
7th Conv	√	128	37.98%	34.35%	41.60%	97.28%	97.01%	98.35%	53.01%	96.9%
10th Conv	√	64	0.00%	0.00%	0.00%	89.78%	91.57%	82.69%	5.08%	53.2%
13th Conv	√	32	0.00%	0.00%	0.00%	81.00%	79.67%	86.26%	0.00%	0.00%

9.3.3 Comparative experiments on multi-line license plates

As shown in Table 9.3, for the dataset RodoSol, we compare our method with the character segmentation method CR-Net, line segmentation method, 1D-sequence recognition methods LPRNet, TRBA, RNN-based segmentation-free method RNN-Attention, and CNN-based segmentation-free method SALPR.

Table 9.3   Comparative experiments on RodoSol

Method	E2E?	UFPR			FPS
		Avg	Single	Double	
SSD+LPRNet		33.24%	65.08%	1.40%	40
SSD(CIoU)+LPRNet		34.23%	66.88%	1.58%	40
Chen et al. +LPRNet		37.23%	72.66%	1.80%	35
CR-Net		55.80%	—	—	—
SSD+Cao et al.		56.35%	65.50%	47.20%	32
SSD(CIoU)+Cao et al.		57.53%	67.20%	47.85%	32

9.3 Experiments

Continued Table 9.3

Method	E2E?	UFPR			FPS
		Avg	Single	Double	
TRBA		59.60%	—	—	—
Chen et al.+Cao et al.		61.55%	69.85%	53.25%	27
SSD+SALPR		91.65%	94.15%	89.15%	36
SSD(CIoU)+SALPR		91.98%	93.95%	90.00%	36
Chen et al.+SALPR		92.04%	93.95%	90.13%	31
SSD+RNN-Attention		92.79%	95.10%	90.48%	33
SSD(CIoU)+RNN-Attention		93.48%	95.78%	91.18%	33
Chen et al.+RNN-Attention		93.84%	95.98%	91.70%	28
Ours small	√	94.01%	95.88%	92.15%	104
Ours large	√	96.55%	98.08%	95.03%	37

The detectors are the vanilla SSD, SSD(CIoU), or Chen et al. SSD(CIoU) denotes using CIoU loss for bounding box regression instead of Smooth L_1 loss in the vanilla SSD. Chen et al. can detect the multidirectional license plate by regressing its vertices. Generally, Chen et al. perform better but slower than SSD and SSD(CIoU). We use SSD(CIoU) as the detector in our end-to-end network. The comparative experiments include two types: a separate detector and recognizer and an end-to-end network. For example, SSD+LPRNet means a separate detector SSD and recognizer LPRNet. Ours Small and Ours Large are our proposed end-to-end networks. With a large backbone, our method can achieve the best recognition performance with a fast inference speed. When using a small backbone, our method can still outperform all comparative methods in recognition and achieve a breakneck inference speed of 104 FPS, proving its effectiveness and efficiency. Notably, LPRNet performs poorly on double-line license plates because character features of different lines are squeezed together, causing mutual interference. Moreover, the character segmentation and line segmentation methods generally lag behind the segmentation-free methods because of character detection errors or inaccurate bisection. However, SALPR cannot effectively perceive plate type and character layout, causing attention errors to affect recognition. RNN-Attention runs slowly due to the step-by-step process.

As shown in Table 9.4, our method achieves the best recognition performance and fastest inference speed on UFPR, verifying its generality. Sighthound, OpenALPR, and Laroca et al. are LPR methods based on character segmentation. These methods perform worse on double-line license plates because of the irregular character layouts. Notably, we do not compare the multidirectional license plate detectoron UFPR because of no vertex annotations.

Table 9.4 Comparative experiments on UFPR

Method	E2E?	UFPR			FPS
		Avg	Single	Double	
SSD+LPRNet		32.11%	37.60%	11.81%	40
SSD(CIoU)+LPRNet		35.33%	42.13%	8.79%	40
Sighthound		47.40%	58.40%	3.30%	—
OpenALPR		50.90%	58.00%	22.80%	—
Laroca et al.		64.90%	72.20%	35.60%	—
TRBA		72.90%	—	—	—
SSD+Cao et al.		74.44%	78.62%	57.97%	32
CR-Net		78.30%	—	—	—
SSD+RNN-Attention		83.56%	87.05%	69.78%	33
SSD+SALPR		83.72%	89.55%	60.71%	36
SSD(CIoU)+Cao et al.		85.83%	86.49%	83.24%	32
Laroca et al.		90.00%	95.90%	66.30%	—
SSD(CIoU)+RNN-Attention		90.00%	90.81%	86.81%	33
SSD(CIoU)+SALPR		93.28%	95.47%	84.62%	36
Ours small	√	94.94%	96.59%	88.46%	104
Ours large	√	99.67%	99.58%	100.00%	37

9.3.4 Comparative experiments on single-line license plates

We test the dataset CCPD to verify the generality of our method, which contains only single-line license plates. As shown in Table 9.5, our method can achieve the best average recognition accuracy and perform best on most subsets of CCPDv2. However, our method performs worse than Chen et al. +RNN-Attention on the subsets Rotate and Tilt that contain highly tilted license plates. Chen et al. can detect multidirectional license plates by vertex regression, while our detector can only detect horizontal license plates. In future work, we will improve the detection and recognition of multidirectional license plates.

Table 9.5 Comparative experiments on CCPDv2

Method	E2E?	Avg	DB	Blur	FN	Rotate	Tilt	Cha.	FPS
SSD+LPRNet		27.74%	27.74%	20.59%	33.57%	20.19%	14.34%	34.57%	40
SSD(CIoU)+LPRNet		28.47%	28.47%	21.49%	34.54%	20.65%	14.48%	35.48%	40
SSD+Cao et al.		28.95%	28.95%	19.63%	32.60%	24.64%	19.38%	33.74%	32
SSD(CIoU)+Cao et al.		29.82%	29.82%	20.63%	34.06%	25.30%	20.09%	34.46%	32
Chen et al.+Cao et al.		41.43%	41.43%	17.51%	40.03%	69.45%	53.50%	34.69%	27
Chen et al.+LPRNet		41.53%	41.53%	18.00%	40.06%	67.87%	52.36%	35.54%	35
SSD+SALPR		43.34%	43.34%	26.67%	45.78%	46.15%	41.18%	45.29%	36
SSD+HC		43.42%	43.42%	25.83%	45.24%	52.82%	52.04%	44.62%	11
SSD(CIoU)+SALPR		44.45%	44.45%	27.87%	47.51%	48.83%	42.29%	46.10%	36
SSD+R-A		47.23%	47.23%	27.84%	48.98%	53.70%	47.40%	47.35%	33
SSD(CIoU)+R-A		47.89%	47.89%	28.82%	49.66%	54.59%	47.71%	48.10%	33
Chen et al.+SALPR		51.91%	51.91%	23.25%	53.13%	82.43%	68.14%	43.54%	31
Chen et al.+R-A		57.23%	57.23%	24.11%	58.89%	88.02%	78.05%	46.66%	28
Ours small	√	70.68%	70.68%	65.35%	68.76%	69.32%	66.42%	87.6%	104
Ours large	√	73.48%	73.48%	63.00%	75.18%	75.70%	73.40%	90.0%	37

Moreover, as shown in Table 9.6, our method can achieve state-of-the-art average recognition accuracy and perform best on most subsets of CCPDv1, significantly outperforming all end-to-end methods. Our method can achieve the fastest inference speed with a small backbone while outperforming most comparative methods in recognition.

Table 9.6 Comparative experiments on CCPDv1

Method	E2E?	Avg	Base	DB	FN	Rotate	Tilt	Wea.	Cha.	FPS
Zherzdev et al.		93.0%	97.8%	92.2%	91.9%	79.4%	85.8%	92.0%	69.8%	56
TE2E	√	94.4%	97.8%	94.8%	94.5%	87.9%	92.1%	86.8%	81.2%	3
Silva et al.		94.6%	98.7%	86.5%	85.2%	94.5%	95.4%	94.8%	91.2%	31
RPNet	√	95.5%	98.5%	96.9%	94.3%	90.8%	92.5%	87.9%	85.1%	61
DAN		96.6%	98.9%	96.1%	96.4%	91.9%	93.7%	95.4%	83.1%	19
MANGO	√	96.9%	99.0%	97.1%	95.5%	95.0%	96.5%	95.9%	83.1%	8
HomoNet	√	97.5%	99.1%	96.9%	95.9%	97.1%	98.0%	97.5%	85.9%	19
ILPRNet		97.5%	99.2%	98.1%	98.5%	90.3%	95.2%	97.8%	86.2%	—
Qin et al.	√	97.5%	99.5%	93.3%	93.7%	98.2%	95.9%	98.9%	92.9%	26
MORAN		98.3%	99.5%	98.1%	98.6%	98.1%	98.6%	97.6%	86.5%	55
AttentionNet		98.5%	99.6%	98.8%	98.8%	96.4%	97.6%	98.5%	88.9%	40
Ours small	√	98.1%	99.3%	98.3%	98.0%	97.3%	97.5%	98.2%	87.6%	104
Ours large	√	98.8%	99.7%	99.1%	98.8%	98.4%	98.5%	98.8%	90.0%	37

9.3.5 Cross-dataset experiments

We conduct cross-dataset experiments on PKUData and CLPD with the training set CCPD-Base. As shown in Table 9.7, our method can achieve the best cross-dataset recognition performance, verifying its generality. PKUData has similar region code distributions with the training set. Hence, the recognition accuracy with and without region codes has little difference. However, CLPD differs from the training set in region code distributions, so the recognition accuracy significantly lags when considering the region code.

Table 9.7　Cross-dataset experiments

Method	PKUData		CLPD	
	w/o R.C.	Overall	w/o R.C.	Overall
Sighthound	89.3%	—	85.2%	—
RPNet	78.4%	77.6%	78.9%	66.5%
AttentionNet	86.5%	84.8%	86.1%	70.8%
AttentionNet w/Syn.	90.5%	88.2%	87.6%	76.8%
Ours	92.8%	92.7%	92.4%	81.1%

Please see colorful images

Chapter 10 Multi-task Learning for License Plate Recognition in Unconstrained Scenarios

The recognition of license plates in natural scenes often face challenges such as multi-directional and multi-line variations. Additionally, previous studies have treated license plate detection and recognition as separate tasks, resulting in inefficiencies and error accumulation. To address these challenges, we propose an end-to-end method for license plate detection and recognition using multi-task learning. Firstly, we introduce two parallel branches to detect the horizontal bounding box and the four corners of the license plate, enabling multi-directional license plate detection in a multi-task manner. The outputs from these branches are combined to enhance recognition accuracy. Secondly, we propose to extract global features to perceive character layout and utilize reading order to spatially attend to characters for recognizing multi-line license plates. Finally, we combine detection and recognition using the same backbone, with the detection branch based on multiple deep layers and the recognition branch based on multiple shallow layers, thereby constructing an end-to-end detection and recognition network. Comparative experiments on CCPD and RodoSol datasets validate that our method significantly outperforms state-of-the-art methods, particularly in scenarios involving multi-directional and multi-line license plates.

10.1 Problem formulation

Automatic License Plate Recognition (ALPR) is a crucial technology in intelligent transportation systems, essential for traffic monitoring, electronic toll collection (ETC), and smart parking lots. While effective in specific operational contexts, ALPR faces challenges in unconstrained scenarios, as shown in Fig. 10.1. Recognizing multi-directional and multi-line license plates poses significant obstacles for conventional ALPR systems.

As depicted in Fig. 10.2(a), previous license plate detectors have shown limitations in effectively handling multi-directional license plates due to their inherent restriction to detecting only the horizontal bounding box. Consequently, this limitation can introduce

Fig. 10.1 **Recognizing license plates in unconstrained scenarios presents challenges**

significant background noises into sub sequent recognition processes, thereby impeding accuracy. To address this constraint, certain studies have proposed a method to predict the four corners of multi-directional license plates, as illustrated in Fig. 10.2 (b). However, when faced with horizontally oriented license plates, detecting the four corners may yield inferior results compared to detecting the bounding box. This discrepancy arises from the inherently more complex nature of corner detection, constituting an 8-DoF(Degree of Freedom) task involving both horizontal and vertical coordinates of the four corners, whereas detecting a horizontal bounding box involves a more straightforward 4-DoF task, considering only the coordinates of the center along with the height and width. To effectively overcome these challenges, our proposed method strategically integrates the strengths of both horizontal and multi-directional detectors. This integration is achieved through a multi-task framework that concurrently detects both the horizontal bounding box and the four corners. By adopting this approach, our method aims to optimize detection accuracy across diverse plate orientations, effectively navigating the intricacies associated with multi-directional license plate recognition.

In contrast to single-line license plates, the recognition of multi-line license plates poses greater challenges due to their inherently complex character layout. As depicted in Fig. 10.2 (c)-(f), previous methodologies for multi-line license plate recognition encompass character segmentation, line segmentation, and segmentation-free methods. Character segmentation methods involve the identification of individual characters followed by the aggregation into character strings. However, these approaches often rely heavily on manual annotation, which can be labor-intensive and prone to errors. Line segmentation techniques horizontally divide the license plate into multiple segments and subsequently concatenate them horizontally. However, horizontal lines may intersect with characters on the license plate, leading to incomplete character

10.1 Problem formulation

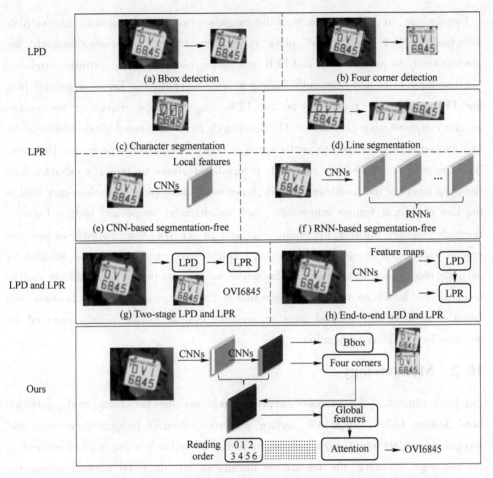

Fig. 10.2 Comparison of different schemes for license plate detection and license plate recognition

recognition. Segmentation-free methods utilize convolutional neural networks (CNNs) or recurrent neural networks (RNNs) to spatially attend to license plate characters without explicit segmentation. While CNN-based methods excel at extracting local features, they may lack a holistic perception of character layout. On the other hand, RNN-based methods leverage their intrinsic characteristics of state computation for character recognition. However, the sequential nature of RNNs makes them challenging to parallelize, resulting in computationally intensive operations and prolonged processing times. Our proposed method leverages local features extracted by CNNs to derive comprehensive global features, enabling a nuanced perception of the character layout. Subsequently, our approach utilizes the reading order to efficiently attend to characters in parallel, thereby optimizing inference time and computational efficiency.

Furthermore, an ALPR system typically comprises two primary subtasks: license plate detection (LPD) and license plate recognition (LPR). Previous research has predominantly focused on LPD and LPR in isolation, overlooking the intrinsic correlation between these tasks and potentially leading to error accumulation. Errors originating from the LPD stage can propagate to the LPR stage, thereby impacting the overall accuracy. Recent advancements in ALPR research have introduced single-stage end-to-end methodologies designed to jointly optimize both LPD and LPR processes. However, it's imperative to acknowledge that while recognition features are typically extracted from the deep layers of the backbone network, repeated downsampling operations may lead to the loss of critical feature information, such as character shape and texture. Previous research proposes utilizing the shallowest layer of the backbone network to preserve crucial features for recognition; however, a single layer may lack diverse features for accurate recognition. To address this limitation, we propose to combine multiple shallow layers of the backbone network for recognition. By incorporating diverse features with larger resolutions, our method aims to preserve crucial feature details essential for accurate license plate recognition.

10.2 Methodology

Fig. 10.3 illustrates the proposed network, including the backbone, neck, detection head, feature fusion, recognition feature, and recognition. It integrates detection and recognition models into a single network trained in an end-to-end approach. Compared to the two-stage approach, the recognition module in our proposed network reuses the features extracted by the detection module as inputs to the LPR. As a result, we do not need to re-extract LP features in the recognition module, which saves significant computational cost. Second, sharing features between the detection and recognition modules strengthens the intrinsic connection between the two tasks. Intuitively, the license plate features contain character features, which are also very important for determining whether it is a license plate or not. Therefore, by integrating both license plate detection and recognition modules into a unified framework, the performance of each module can be further improved.

First, we use CSPDarknet as the backbone to extract shared features, which are used by the neck and the recognizer for feature fusion, respectively, to be used as inputs for the two subsequent tasks. Then, the detection head accurately detects the bounding box and the four corners of the license plate using the features from PAN. For the license plate with bounding box, RoIAlign is used to crop and resize the shared feature map, and

10.2 Methodology

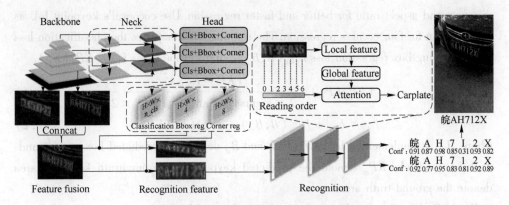

Fig. 10.3 Overall structure of our network

for the license plate with four corners, the perspective transform is used to correct and resize the shared feature map. Finally, the license plate characters will be recognized by recognizer.

We use CSPDarknet as the backbone to extract shared features for detection and recognition. Specifically, five feature maps with different scale sizes (the input size of the backbone is 640×640) are generated by five convolutions each containing Conv2d (kernel=3, stride=2), bn and silu. Then we use PAN to connect them into a path by up-sampling the feature maps starting from the low-resolution ones, while down-sampling them starting from the high-resolution ones. In this process, the information of each layer of feature maps will be fused with the feature maps of the neighboring layers above and below, but unlike FPN, PAN will cascade the results of the fused feature maps of different layers instead of summing them up. This avoids the loss of information in the summation process, thus improving the detection.

In order to solve the problem of multi-directional license plate recognition, we design an anchor-free detection head that predicts the four corners and the bounding box of the license plate at the same time, as shown in Fig. 10.3. There are three sub-branches, including bounding box regression, four corners regression, the last is classification and conference of the license plate. The bounding box information of the license plate will be combined with the shared features of the license plate to crop and resize the shared feature graph through RoIAlign. The corners information of the license plate will be corrected and the size of the shared feature map will be adjusted by perspective transformation.

We use BCEloss as the loss function for classification, and CIoU as the loss function for Bbox regression because the CIoU loss considers the overlap region, central point distance, and aspect ratio for better and faster regression. Use cocoeval's keypoint IoU as loss function for four corners regression. The training loss comprises the classification loss L_{cls}, bounding box regression loss $L_{CIoU}(B, B_{gt})$, and keypoint loss $L_{kpIoU}(KP, KP_{gt})$.

$$L_{kpIoU}(KP, KP_{gt}) = \frac{(KP - KP_{gt})^2}{2 \times \text{sigmas} \times (\text{area} + 10^{-9}) \times 2} \quad (10.1)$$

$$L_{det} = L_{cls} + L_{CIoU}(B, B_{gt}) + L_{kpIoU}(KP, KP_{gt}) \quad (10.2)$$

where, L_{cls} is implemented by BCEloss. B and B_{gt} denote the predicted box and ground-truth box, KP and KP_{gt} denote the predicted keypoint and groundtruth keypoint, area denote the ground-truth area.

For LPD, deep features have stronger semantics and larger receptive fields, which can better help detect the target location. However, deep features are not favorable for LPR because the character features needed for LPR belong to small targets, such as character shapes and textures, which lose a lot of information after multiple down-sampling operations. Whereas shallow features are semantically weak and retain a lot of information about the characters. In our backbone, the first convolutional layer and the second convolutional layer still retain a large amount of character information from two different scales. Therefore, we fuse two feature layers as recognition features to further improve the recognition performance.

We resize the recognition features to 96×32 as input for the recognizer. Our recognizer consist of local feature extraction, global feature generation, and parallel character attention. Specifically, we first obtain 24×8 local features through multiple convolutional layers. Then, we stack two transformers units to extract global features, as shown in Fig. 10.4. Because local features cannot perceive the type of license plate and character layout, while global features can better enhance character attention. In sequence recognition, attention is used to generate N features, each corresponding to a character in the text. Existing methods typically use the hidden state H_{t-1} as a query to generate the t-th feature. To enable parallel computation, we encode the character reading order by parallel embedding as the query, while the key-value set is obtained from the global features. Finally, we use a fully-connected layer to map to class probabilities $P_n \times cls$, where n represents the number of characters in the license plate, cls represents the number of character categories. We use cross-entropy loss as the recognition loss to maximize the prediction confidence of each character.

10.3 Experiments

We evaluate our proposed end-to-end ALPR model through experiments on several

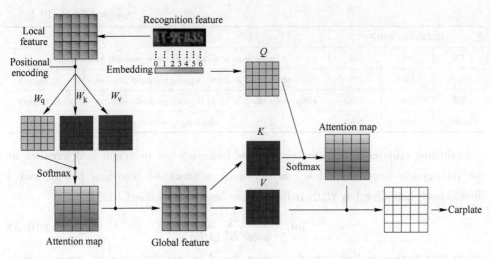

Fig. 10. 4 The architecture of recognition network

datasets.

Dataset CCPD is the Chinese City Parking Dataset, which includes two versions, CCPDv1 and CCPDv2. Both versions comprise over 250000 car images with a resolution of 720×1160 pixels. Each license plate in the dataset is a single-line license plate. The detailed information of the CCPD dataset is presented in the table. CCPDv1 is divided into subsets: Base, DB, FN, Rotate, Tilt, Challenge, and Weather. CCPDv2 excludes the Weather subset but includes an additional Blur subset. The Base subset is used for training and validation, while the other subsets are used for testing. The test set of CCPDv2 is more diverse and challenging compared to CCPDv1. Table 10. 1 provides an overview of the characteristics of the CCPD dataset.

RodoSol is collected at toll stations on highways in Brazil and consists of 20000 car images with a resolution of 1280×720 pixels. The dataset is divided into training, testing, and validation sets according to a 4 : 4 : 2 ratio, with an equal number of single-line and double-line license plates in each set. Notably, in all the mentioned datasets, each image contains only one license plate.

Table 10. 1 Description of the CCPD dataset, including CCPDv1 and CCPDv2

	CCPDv1	-CCPDv2	Description
Base	200k	200k	Normal license plate
DB	20k	10k	License plate under low or high lighting conditions
Blur	—	20k	Blur license plate
Weather	20k	—	License plate in rainy, snowy, or foggy weather

Continued Table 10.1

CCPDv1-CCPDv2			Description
FN	10k	20k	License plate at a distant or close proximity to the camera
Rotate	10k	10k	Horizontally tilted at 20-50 degrees, vertically tilted at −10 to 10 degrees
Tilt	10k	30k	Horizontally tilted at 15-45 degrees, vertically tilted at −15 to 45 degrees
Challenge	10k	50k	Challenging license plate

Evaluation criterion Firstly, the accuracy of bounding box detection is determined by the Intersection-over-Union (IoU) metric. Given a predicted bounding box B' and a ground truth bounding box B, their IoU is calculated as Equation(10.3).

$$\text{IoU} = \frac{\text{area}(B' \cap B)}{\text{area}(B' \cup B)} \quad (10.3)$$

where area represents the area of a region. Based on existing research, we consider a detection to be accurate only if IoU > 0.7. Subsequently, a license plate is considered correctly recognized only when all predicted characters match the ground truth values. We use accuracy to measure the effectiveness of different algorithms. *TP* represents samples that are correctly predicted. *FP* represents samples that are incorrectly predicted. The accuracy of LPD/LPR algorithms are defined as Equation(10.4).

$$\text{accuracy} = \frac{TP}{TP + FP} \quad (10.4)$$

We test the dataset CCPD to verify the generality of our method. As shown in Table 10.2, we compare the two-stage method and end-to-end method, where the detectors of the two-stage method include SSD, SSD (CIoU), and Chen et al., and the recognizers include the line segmentation method, the 1D-sequence recognition methods LPRNet, RNN-based segmentation-free method RNN-Attention, and CNN-based segmentation-free method SALPR. Our method can achieve the best recognition performance with a fast inference speed on all subsets of CCPDv2, proved that our method can effectively recognize multi-directional license plates and horizontal license plates.

Moreover, as shown in Table 10.3, our method can achieve state-of-the-art average recognition accuracy, significantly outperforming all end-to-end methods. However, our method does not achieve the best inference speed because our method spends more time in recognizing the two branches of the detected head. In future work, we will further optimize the recognition time of our method.

We compared the effectiveness of double-line license plate recognition on the RodoSol dataset, as shown in Table 10.4. In the comparative method, the detectors for the two-

Table 10.2 Comparative experiments on CCPDv2

Method	E2E	Avg	DB	Blur	FN	Rotate	Tilt	Cha.	FPS
SSD+LPRNet		27.74%	21.59%	20.59%	33.57%	20.19%	14.34%	34.57%	40
SSD(CIoU)+LPRNet		28.47%	22.45%	21.49%	34.50%	20.65%	14.48%	35.48%	40
SSD+Cao et al.		28.95%	24.05%	19.63%	32.60%	24.64%	19.38%	33.74%	32
SSD(CIoU)+Cao et al.		29.82%	25.32%	20.63%	34.06%	25.30%	20.09%	34.46%	32
Chen et al.+Cao et al.		41.43%	26.55%	17.51%	40.03%	69.45%	53.50%	34.69%	27
Chen et al.+LPRNet		41.53%	24.05%	18.00%	40.06%	67.87%	52.36%	35.54%	35
SSD+SALPR		43.34%	35.50%	26.67%	45.78%	46.15%	41.18%	45.29%	36
SSD+HC		43.42%	34.47%	25.83%	45.24%	52.82%	52.04%	44.62%	11
SSD(CIoU)+SALPR		44.45%	36.85%	27.87%	47.51%	48.83%	42.29%	46.10%	36
SSD+RNN-Attention		47.23%	39.66%	27.84%	48.98%	53.70%	47.40%	47.35%	33
SSD(CIoU)+RNN-Attention		47.89%	40.72%	28.82%	49.66%	54.59%	47.71%	48.10%	33
Chen et al.+SALPR		51.91%	34.78%	23.25%	53.13%	82.43%	68.14%	43.54%	31
Chen et al.+RNN-Attention		57.23%	38.81%	24.11%	58.89%	88.02%	78.05%	46.66%	28
Chen et al.	√	73.48%	63.85%	63.00%	75.18%	75.70%	73.40%	74.63%	37
Ours	√	**88.75%**	**84.90%**	**80.27%**	**92.20%**	**97.17%**	**94.04%**	**86.68%**	**41**

Table 10.3 Comparative experiments on CCPDv1

Method	E2E	Avg	DB	Blur	FN	Rotate	Tilt	Cha.	FPS
LPRNet		93.0%	92.2%	91.9%	79.4%	85.8%	92.0%	69.8%	56
TE2E	√	94.4%	94.8%	94.5%	87.9%	92.1%	86.8%	81.2%	3
Silva et al.		94.6%	86.5%	85.2%	94.5%	95.4%	94.8%	91.2%	31
RPNet	√	95.5%	96.9%	94.3%	90.8%	92.5%	87.9%	85.1%	**61**
DAN		96.6%	96.1%	96.4%	91.9%	93.7%	95.4%	83.1%	19
MANGO	√	96.9%	97.1%	95.5%	95.0%	96.5%	95.9%	83.1%	8
HomoNet	√	97.5%	96.9%	95.9%	97.1%	98.0%	97.5%	85.9%	19
ILPRNet		97.5%	98.1%	98.5%	90.3%	95.2%	97.8%	86.2%	—
Qin et al.	√	97.5%	93.3%	93.7%	98.2%	95.9%	98.9%	92.9%	26
MORAN		98.3%	98.1%	98.6%	98.1%	98.6%	97.6%	86.5%	55
AttentionNet		98.5%	98.8%	98.8%	96.4%	97.6%	98.5%	88.9%	40
Chen et al.	√	98.8%	99.1%	98.8%	98.4%	98.5%	98.8%	90.0%	37
Ours	√	**99.1%**	**99.7%**	**99.5%**	**99.7%**	**99.7%**	**99.8%**	**95.1%**	41

Table 10.4 Comparative experiments on RodoSol

Method	E2E	RodoSol			FPS
		Avg	Single	Double	
SSD+SALPR		91.65%	94.15%	89.15%	36
SSD(CIoU)+SALPR		91.98%	93.95%	90.00%	36
Chen et al.+SALPR		92.04%	93.95%	90.13%	31
SSD+RNN-Attention		92.79%	95.10%	90.48%	33
SSD(CIoU)+RNN-Attention		93.48%	95.78%	91.18%	33
Chen et al.+RNN-Attention		93.84%	95.98%	91.70%	28
Ours	√	**97.06%**	**97.78%**	**96.35%**	**41**

stage method including vanilla SSD, SSD(CIoU) and Chen et al., SSD(CIoU) denotes using CIoU loss for bounding box regression. The recognizer for the two-stage method including CNN-based segmentation-free method SALPR and RNN-based segmentation-free method RNN-Attention. Our method obtains a significant improvement in double-line license plate recognition performance.

Ablation studies on recognizer components as shown in Table 10.5, we conduct ablation experiments of different recognizer components on CCPD and RodoSol, including global feature generation(GFG) and parallel character attention(PCA). After removing GFG, our proposed method uses local features instead of global features as input to PCA, which significantly reduces the recognition accuracy, especially for CCPDv2 and double-line license plate of Rodosol, however, inference speed almost unchanged. This suggests that global features can better enhance character attention and perceive character layout, thus improve recognition accuracy. Without PCA, our proposed method will degrade to the RNN based segmentation-free model. RNN introduces temporal information of sequences, however, there is no or minor sequential relationship between license plate characters, which as noise affects the recognition accuracy. Moreover, RNN can only encode sequentially, which is time-consuming.

Table 10.5 Ablation study of different recognizer components

Input	CCPDv1	CCPDv2	RodoSol			FPS
			Avg	Single	Double	
Ours	**99.17%**	**88.75%**	**97.06%**	**97.78%**	**96.35%**	**41**
w/o GFG	98.91%	84.17%	95.71%	97.27%	94.15%	41
w/o PCA	98.65%	81.31%	94.58%	96.57%	92.60%	37

The impact of recognition features on LPR The choice of different recognition features significantly impacts the accuracy of the LPR module, especially as deep features lose a considerable amount of character information, which is detrimental to LPR. We conducted tests on three datasets to assess the impact of selecting different recognition features as inputs for the recognizer. As shown in Fig. 10.5, "Image" represents a non-end-to-end two-stage detection and recognition, using the same detector and recognizer as the end-to-end network. This means that the input to the recognizer is the original image. "Layer x" denotes selecting the feature map output of the x-th convolutional layer in the backbone as the input for the recognizer. Performance was notably enhanced on all three datasets when choosing the first feature map as the recognition feature, compared to the two-stage detection and recognition. This improvement is attributed to the joint optimization network, which reduces error accumulation and improves recognition rates, while shallow features retain abundant character information. However, selecting deeper features as recognition features led to a significant decrease in recognition accuracy. This is because deep features lose character information after multiple down-sampling operations, negatively impacting LPR.

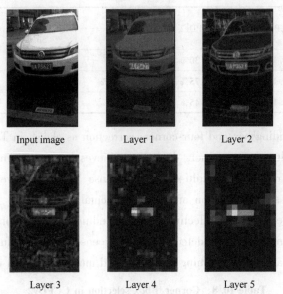

Fig. 10.5 Feature visualization of the input image and different convolutional layers

As shown in Table 10.6, selecting the first feature map from backbone achieved the best recognition performance across all datasets.

Table 10.6 Ablation study of different convolutional layers as recognition inputs

Input	Size	CCPDv1	CCPDv2	RodoSol		
				Avg	Single	Double
Image		98.76%	84.90%	96.73%	97.65%	95.80%
Layer 1	320	**99.17%**	**88.30%**	96.93%	97.73%	**96.13%**
Layer 2	160	97.69%	83.38%	83.21%	84.65%	81.76%
Layer 3	80	87.28%	54.62%	3.25%	3.75%	2.75%
Layer 4	40	60.17%	23.01%	0%	0%	0%
Layer 5	20	10.48%	3.01%	0%	0%	0%

However, when choosing the second feature map as the recognition feature, recognition accuracy did not decrease signiffcantly, indicating that the second feature map still retains some larger-scale information, such as character layout. Therefore, we propose concat the first and second feature maps as the recognition feature, encompassing feature information at different scales and further enhancing the recognition accuracy. As shown in Table 10.7, experimental results demonstrate that we achieved the highest recognition accuracy when using the first and second feature maps for feature fusion.

Table 10.7 Ablation study of feature fusion

Method	CCPDv1	CCPDv2	RodoSol		
			Avg	Single	Double
Layer 1	99.17%	88.30%	96.93%	97.73%	96.13%
Layers 1+2	99.17%	**88.75%**	**97.06%**	**97.78%**	**96.35%**
Layers 1+2+3	**99.18%**	88.45%	95.14%	96.38%	93.90%

Selection of bounding box and four-corners correction as shown in Table 10.8, in the FN, Rotate, and Tilt subsets, corners detection achieves better performance, addressing the challenges of recognizing multi-direction license plates. However, on subsets of license plates with smaller rotation angles or horizontal, such as DB, Blur, Challenge, corners detection performs less effectively than bounding box detection. Proved that the corner detector cannot effectively detect horizontal license plates. We further improve the recognition performance by combining horizontal and multi-directional detectors.

Table 10.8 Corner/Bbox selection in CCPDv2

Method	Avg	DB	Blur	FN	Rotate	Tilt	Cha.	FPS
Corner	85.84%	81.45%	75.66%	**89.09%**	**96.46%**	**91.61%**	83.95%	53
Bbox	86.14%	83.13%	78.76%	88.61%	96.17%	88.73%	85.18%	**75**
C/B selection	**88.75%**	**84.90%**	**80.27%**	**92.20%**	**97.17%**	**94.04%**	**86.68%**	41

10.3 Experiments

We validated the generalizability of our approach on the CCPDv1, CCPDv2, and RodoSol. As shown in Table 10.9, the results indicate that the selection method performs the best across all three datasets.

Table 10.9 Ablation study of corner/Bbox selection

Method	CCPDv1		CCPDv2		RodoSol	
	Layer 1	Layers 1+2	Layer 1	Layers 1+2	Layer 1	Layers 1+2
Corner	99.02%	98.99%	85.62%	85.84%	95.74%	96.46%
Bbox	99.04%	98.94%	86.47%	86.14%	95.95%	95.83%
C/B selection	**99.17%**	**99.17%**	**88.30%**	**88.75%**	**96.93%**	**97.06%**

Please see colorful images

References

[1] Redmon J. You only look once: Unified, real-time object detection[C]//Proceedings of the IEEE Conference on Computer Vision and Pattern Recognition. 2016.

[2] Ren S, He K, Girshick R, et al. Faster R-CNN: Towards real-time object detection with region proposal networks[J]. IEEE Transactions on Pattern Analysis and Machine Intelligence, 2016, 39(6):1137-1149.

[3] Zeiler M D, Fergus R. Visualizing and understanding convolutional networks [C]//Computer Vision-ECCV 2014: 13th European Conference, Zurich, Switzerland, September 6-12, 2014, Proceedings, Part I 13. Springer International Publishing, 2014:818-833.

[4] Lin T Y, Maji S. Visualizing and understanding deep texture representations[C] //Proceedings of the IEEE Conference on Computer Vision and Pattern Recognition. 2016:2791-2799.

[5] Redmon J, Farhadi A. YOLO9000: Better, faster, stronger [C]//Proceedings of the IEEE Conference on Computer Vision and Pattern Recognition. 2017:7263-7271.

[6] Redmon J. Yolov3:An incremental improvement[J]. arXiv preprint arXiv:1804.02767, 2018.

[7] Wang J, Yuan Y, Yu G. Face attention network: An effective face detector for the occluded faces [J]. arXiv preprint arXiv:1711.07246, 2017.

[8] Felzenszwalb P F, Girshick R B, McAllester D, et al. Object detection with discriminatively trained part-based models[J]. IEEE Transactions on Pattern Analysis and Machine Intelligence, 2009, 32(9):1627-1645.

[9] Girshick R, Donahue J, Darrell T, et al. Rich feature hierarchies for accurate object detection and semantic segmentation[C]//Proceedings of the IEEE Conference on Computer Vision and Pattern Recognition. 2014:580-587.

[10] Uijlings J R R, Van De Sande K E A, Gevers T, et al. Selective search for object recognition [J]. International journal of computer vision, 2013, 104:154-171.

[11] Girshick R. Fast R-CNN[C]//Proceedings of the IEEE International Conference on Computer Vision(ICCV). 2015:1440-1448.

[12] Lin T Y, Dollár P, Girshick R, et al. Feature pyramid networks for object detection [C]// Proceedings of the IEEE Conference on Computer Vision and Pattern Recognition. 2017: 2117-2125.

[13] Russakovsky O, Deng J, Su H, et al. Imagenet large scale visual recognition challenge [J]. International Journal of Computer Vision, 2015, 115:211-252.

[14] Diederik P K. Adam:A method for stochastic optimization[J]. (No Title), 2014.

[15] Kuznetsova A, Rom H, Alldrin N, et al. The open images dataset v4: Unified image classification, object detection, and visual relationship detection at scale[J]. International Journal of Computer Vision, 2020, 128(7):1956-1981.

[16] Krause J, Stark M, Deng J, et al. 3d object representations for fine-grained categorization[C]// Proceedings of the IEEE International Conference on Computer Vision Workshops. 2013: 554-561.

[17] Hsu G S, Chen J C, Chung Y Z. Application-oriented license plate recognition[J]. IEEE Transactions on Vehicular Technology,2012,62(2):552-561.

[18] Everingham M,Van Gool L,Williams C K I,et al. The pascal visual object classes(voc)challenge [J]. International Journal of Computer Vision,2010,88:303-338.

[19] Li Y,Chen Y,Wang N,et al. Scale-aware trident networks for object detection[C]//Proceedings of the IEEE/CVF International Conference on Computer Vision. 2019:6054-6063.

[20] Yan J, Cho M, Zha H, et al. Multi-graph matching via affinity optimization with graduated consistency regularization[J]. IEEE Transactions on Pattern Analysis and Machine Intelligence, 2015,38(6):1228-1242.

[21] Yan J, Li C, Li Y, et al. Adaptive discrete hypergraph matching[J]. IEEE Transactions on Cybernetics,2017,48(2):765-779.

[22] Ge S, Zhao S, Li C, et al. Low-resolution face recognition in the wild via selective knowledge distillation[J]. IEEE Transactions on Image Processing,2018,28(4):2051-2062.

[23] Tian Y,Song J,Zhang X,et al. An algorithm combined with color differential models for license-plate location[J]. Neurocomputing,2016,212:22-35.

[24] Zhou W,Li H,Lu Y,et al. Principal visual word discovery for automatic license plate detection [J]. IEEE Transactions on Image Processing,2012,21(9):4269-4279.

[25] Li B,Tian B,Li Y,et al. Component-based license plate detection using conditional random field model[J]. IEEE Transactions on Intelligent Transportation Systems,2013,14(4):1690-1699.

[26] Al-Ghaili A M, Mashohor S, Ramli A R, et al. Vertical-edge-based car-license-plate detection method[J]. IEEE Transactions on Vehicular Technology,2012,62(1):26-38.

[27] Han J,Yao J,Zhao J,et al. Multi-oriented and scale-invariant license plate detection based on convolutional neural networks[J]. Sensors,2019,19(5):1175.

[28] Yuan Y, Zou W, Zhao Y, et al. A robust and efficient approach to license plate detection [J]. IEEE Transactions on Image Processing,2016,26(3):1102-1114.

[29] Al-Qudah R, Suen C Y. Enhancing YOLO deep networks for the detection of license plates in complex scenes[C]//Proceedings of the Second International Conference on Data Science, E-Learning and Information Systems. 2019:1-6.

[30] Pu D, Gu N, Zhang X. A robust and real-time approach for license plate detection[C]//2018 14th International Conference on Natural Computation, Fuzzy Systems and Knowledge Discovery(ICNC-FSKD). IEEE,2018:1-7.

[31] Chen S L, Yang C, Ma J W, et al. Simultaneous end-to-end vehicle and license plate detection with multi-branch attention neural network[J]. IEEE Transactions on Intelligent Transportation Systems,2019,21(9):3686-3695.

[32] Xu Z, Yang W, Meng A, et al. Towards end-to-end license plate detection and recognition:A large dataset and baseline [C]//Proceedings of the European Conference on Computer Vision (ECCV). 2018:255-271.

[33] Li H,Wang P,Shen C. Toward end-to-end car license plate detection and recognition with deep neural networks[J]. IEEE Transactions on Intelligent Transportation Systems, 2018, 20(3):

1126-1136.

[34] Kim S G, Jeon H G, Koo H I. Deep-learning-based license plate detection method using vehicle region extraction[J]. Electronics Letters, 2017, 53(15):1034-1036.

[35] Fu Q, Shen Y, Guo Z. License plate detection using deep cascaded convolutional neural networks in complex scenes[C]//Neural Information Processing: 24th International Conference, ICONIP 2017, Guangzhou, China, November 14-18, 2017, Proceedings, Part II 24. Springer International Publishing, 2017:696-706.

[36] Silva S M, Jung C R. License plate detection and recognition in unconstrained scenarios[C]// Proceedings of the European Conference on Computer Vision(ECCV). 2018:580-596.

[37] Yonetsu S, Iwamoto Y, Chen Y W. Two-stage YOLOv2 for accurate license-plate detection in complex scenes[C]//2019 IEEE International Conference on Consumer Electronics (ICCE). IEEE, 2019:1-4.

[38] Laroca R, Severo E, Zanlorensi L A, et al. A robust real-time automatic license plate recognition based on the YOLO detector[C]//2018 International Joint Conference on Neural Networks (IJCNN). IEEE, 2018:1-10.

[39] Chen S L, Tian S, Ma J W, et al. End-to-end trainable network for degraded license plate detection via vehicle-plate relation mining[J]. Neurocomputing, 2021, 446:1-10.

[40] Montazzolli S, Jung C. Real-time brazilian license plate detection and recognition using deep convolutional neural networks[C]//2017 30th SIBGRAPI Conference on Graphics, Patterns and Images(SIBGRAPI). IEEE, 2017:55-62.

[41] Jin L, Xian H, Bie J, et al. License plate recognition algorithm for passenger cars in Chinese residential areas[J]. Sensors, 2012, 12(6):8355-8370.

[42] Ullah F, Anwar H, Shahzadi I, et al. Barrier access control using sensors platform and vehicle license plate characters recognition[J]. Sensors, 2019, 19(13):3015.

[43] Dalarmelina N V, Teixeira M A, Meneguette R I. A real-time automatic plate recognition system based on optical character recognition and wireless sensor networks for ITS[J]. Sensors, 2019, 20(1):55.

[44] Islam K T, Raj R G, Shamsul Islam S M, et al. A vision-based machine learning method for barrier access control using vehicle license plate authentication[J]. Sensors, 2020, 20(12):3578.

[45] Dong M, He D, Luo C, et al. A CNN-based approach for automatic license plate recognition in the wild[C]//BMVC. 2017.

[46] Dai J, He K, Sun J. Instance-aware semantic segmentation via multi-task network cascades[C]// Proceedings of the IEEE Conference on Computer Vision and Pattern Recognition. 2016: 3150-3158.

[47] Liu W, Anguelov D, Erhan D, et al. Ssd: Single shot multibox detector[C]//Computer Vision-ECCV 2016: 14th European Conference, Amsterdam, The Netherlands, October 11-14, 2016, Proceedings, Part I 14. Springer International Publishing, 2016:21-37.

[48] Simonyan K, Zisserman A. Very deep convolutional networks for large-scale image recognition [C]//Proceedings of the 3rd International Conference on Learning Representations (ICLR),

2015:7-9.

[49] Glorot X, Bengio Y. Understanding the difficulty of training deep feedforward neural networks [C]//Proceedings of the Thirteenth International Conference on Artificial Intelligence and Statistics. JMLR Workshop and Conference Proceedings. 2010:249-256.

[50] Kinga D, Adam J B. A method for stochastic optimization [C]//International Conference on Learning Representations(ICLR). 2015,5:6.

[51] Liao M, Shi B, Bai X, et al. Textboxes: A fast text detector with a single deep neural network [C]//Proceedings of the AAAI Conference on Artificial Intelligence. 2017,31(1):4161-4167.

[52] Batra P, Hussain I, Ahad M A, et al. A novel memory and time-efficient ALPR system based on YOLOv5[J]. Sensors, 2022, 22(14):5283.

[53] Carion N, Massa F, Synnaeve G, et al. End-to-end object detection with transformers [C]// European Conference on Computer Vision. Cham: Springer International Publishing. 2020: 213-229.

[54] Chen S L, Liu Q, Ma J W, et al. Scale-invariant multidirectional license plate detection with the network combining indirect and direct branches[J]. Sensors, 2021, 21(4):1074.

[55] Gonçalves G R, da Silva S P G, Menotti D, et al. Benchmark for license plate character segmentation[J]. Journal of Electronic Imaging, 2016, 25(5):053034.

[56] He K, Zhang X, Ren S, et al. Deep residual learning for image recognition[C]//Proceedings of the IEEE Conference on Computer Vision and Pattern Recognition. 2016:770-778.

[57] Kuhn H W. The Hungarian method for the assignment problem [J]. Naval Research Logistics Quarterly, 1955, 2(1/2):83-97.

[58] Laroca R, Zanlorensi L A, Gonçalves G R, et al. An efficient and layout-independent automatic license plate recognition system based on the YOLO detector [J]. IET Intelligent Transport Systems, 2021, 15(4):483-503.

[59] Lin T Y, Maire M, Belongie S, et al. Microsoft coco: Common objects in context[C]//Computer Vision-ECCV 2014: 13th European Conference, Zurich, Switzerland, September 6-12, 2014, Proceedings, Part V 13. Springer International Publishing, 2014:740-755.

[60] Rezatofighi H, Tsoi N, Gwak J Y, et al. Generalized intersection over union: A metric and a loss for bounding box regression[C]//Proceedings of the IEEE/CVF Conference on Computer Vision and Pattern Recognition. 2019:658-666.

[61] Vaswani A. Attention is all you need [C]//Advances in Neural Information Processing Systems. 2017:5998-6008.

[62] Zou Y, Zhang Y, Yan J, et al. License plate detection and recognition based on YOLOv3 and ILPRNET[J]. Signal, Image and Video Processing, 2022, 16(2):473-480.

[63] Baek J, Kim G, Lee J, et al. What is wrong with scene text recognition model comparisons? dataset and model analysis[C]//Proceedings of the IEEE/CVF International Conference on Computer Vision. 2019:4715-4723.

[64] Bahdanau D, Cho K, Bengio Y. Neural machine translation by jointly learning to align and translate [C]//Proceedings of the International Conference on Learning Representations

(ICLR), 2015.

[65] Duan S, Hu W, Li R, et al. Attention enhanced ConvNet-RNN for Chinese vehicle license plate recognition[C]//Pattern Recognition and Computer Vision: First Chinese Conference, PRCV 2018, Guangzhou, China, November 23-26, 2018, Proceedings, Part II 1. Springer International Publishing. 2018:417-428.

[66] Gou C, Wang K, Yao Y, et al. Vehicle license plate recognition based on extremal regions and restricted Boltzmann machines [J]. IEEE Transactions on Intelligent Transportation Systems, 2015, 17(4):1096-1107.

[67] Graves A, Fernández S, Gomez F, et al. Connectionist temporal classification: Labelling unsegmented sequence data with recurrent neural networks [C]//Proceedings of the 23rd International Conference on Machine Learning. 2006:369-376.

[68] Huang Y, Luo C, Jin L, et al. Attention after attention: Reading text in the wild with cross attention [C]//2019 International Conference on Document Analysis and Recognition (ICDAR). IEEE. 2019:274-280.

[69] Kessentini Y, Besbes M D, Ammar S, et al. A two-stage deep neural network for multi-norm license plate detection and recognition [J]. Expert Systems with Applications, 2019, 136: 159-170.

[70] Li H, Wang P, Shen C, et al. Show, attend and read: A simple and strong baseline for irregular text recognition[C]//Proceedings of the AAAI Conference on Artificial Intelligence. 2019, 33(1): 8610-8617.

[71] Liu X C, Ma H D, Li S Q. PVSS: A progressive vehicle search system for video surveillance networks[J]. Journal of Computer Science and Technology, 2019, 34:634-644.

[72] Lu N, Yang W, Meng A, et al. Automatic recognition for arbitrarily tilted license plate[C]// Proceedings of the 2018 2nd International Conference on Video and Image Processing. 2018: 23-28.

[73] Luo C, Jin L, Sun Z. Moran: A multi-object rectified attention network for scene text recognition [J]. Pattern Recognition, 2019, 90:109-118.

[74] Ly N T, Nguyen C T, Nakagawa M. An attention-based end-to-end model for multiple text lines recognition in japanese historical documents[C]//2019 International Conference on Document Analysis and Recognition(ICDAR). IEEE. 2019:629-634.

[75] Martínez-Carballido J, Alfonso-López R, Ramírez-Cortés J M. License plate digit recognition using 7×5 binary templates at an outdoor parking lot entrance [C]//CONIELECOMP 2011, 21st International Conference on Electrical Communications and Computers. IEEE. 2011:18-21.

[76] Shao W, Chen L. License plate recognition data-based traffic volume estimation using collaborative tensor decomposition [J]. IEEE Transactions on Intelligent Transportation Systems, 2018, 19(11):3439-3448.

[77] Špaňhel J, Sochor J, Juránek R, et al. Holistic recognition of low quality license plates by CNN using track annotated data[C]//2017 14th IEEE International Conference on Advanced Video and Signal Based Surveillance(AVSS). IEEE. 2017:1-6.

[78] Wang J, Huang H, Qian X, et al. Sequence recognition of Chinese license plates [J]. Neurocomputing, 2018, 317: 149-158.

[79] Wang T, Zhu Y, Cao L, et al. Decoupled attention network for text recognition[C]//Proceedings of the AAAI Conference on Artificial Intelligence. 2020, 34(7): 12216-12224.

[80] Zhang L, Wang P, Li H, et al. A robust attentional framework for license plate recognition in the wild[J]. IEEE Transactions on Intelligent Transportation Systems, 2020, 22(11): 6967-6976.

[81] Zherzdev S, Gruzdev A. LPRNet: License plate recognition via deep neural networks[J]. arXiv preprint arXiv: 1806. 10447, 2018.

[82] Zhang L, Wang P, Dang F, et al. A simple and robust attentional encoder-decoder model for license plate recognition [C]//Pattern Recognition and Computer Vision: Second Chinese Conference, PRCV 2019, Xi'an, China, November 8-11, 2019, Proceedings, Part I 2. Springer International Publishing. 2019: 295-307.

[83] Laroca R, Cardoso E V, Lucio D R, et al. On the cross-dataset generalization in license plate recognition[J]. arXiv preprint arXiv: 2201. 00267, 2022.

[84] Silva S M, Jung C R. Real-time license plate detection and recognition using deep convolutional neural networks [J]. Journal of Visual Communication and Image Representation, 2020, 71: 102773.

[85] Li H. Reading Car License Plates Using Deep Convolutional Neural Networks and LSTMs [J]. arXiv preprint arXiv: 1601. 05610, 2016.

[86] Liu Q, Chen S L, Li Z J, et al. Fast recognition for multidirectional and multi-type license plates with 2D spatial attention [C]//Document Analysis and Recognition-ICDAR 2021: 16th International Conference, Lausanne, Switzerland, September 5-10, 2021, Proceedings, Part IV 16. Springer International Publishing. 2021: 125-139.

[87] Xu H, Zhou X D, Li Z, et al. EILPR: Toward end-to-end irregular license plate recognition based on automatic perspective alignment[J]. IEEE Transactions on Intelligent Transportation Systems, 2021, 23(3): 2586-2595.

[88] Yu D, Li X, Zhang C, et al. Towards accurate scene text recognition with semantic reasoning networks [C]//Proceedings of the IEEE/CVF conference on computer vision and pattern recognition. 2020: 12113-12122.

[89] Wang H, Qian Y, Wang X, et al. Improving noise robustness of contrastive speech representation learning with speech reconstruction[C]//ICASSP 2022-2022 IEEE International Conference on Acoustics, Speech and Signal Processing(ICASSP). IEEE. 2022: 6062-6066.

[90] Khorram S, Kim J, Tripathi A, et al. Contrastive siamese network for semi-supervised speech recognition[C]//ICASSP 2022-2022 IEEE International Conference on Acoustics, Speech and Signal Processing(ICASSP). IEEE. 2022: 7207-7211.

[91] Han X, Luo Y, Chen W, et al. Cross-lingual contrastive learning for fine-grained entity typing for low-resource languages[C]//Proceedings of the 60th Annual Meeting of the Association for Computational Linguistics(Volume 1: Long Papers). 2022: 2241-2250.

[92] Cui W, Zheng G, Wang W. Unsupervised natural language inference via decoupled multimodal

contrastive learning[J]. arXiv preprint arXiv:2010. 08200,2020.

[93] Hjelm R D, Fedorov A, Lavoie-Marchildon S, et al. Learning deep representations by mutual information estimation and maximization[J]. arXiv preprint arXiv:1808. 06670,2018.

[94] He K, Fan H, Wu Y, et al. Momentum contrast for unsupervised visual representation learning [C]//Proceedings of the IEEE/CVF Conference on Computer Vision and Pattern Recognition. 2020:9729-9738.

[95] Chen T, Kornblith S, Norouzi M, et al. A simple framework for contrastive learning of visual representations[C]//International Conference on Machine Learning. PMLR. 2020:1597-1607.

[96] Khosla P, Teterwak P, Wang C, et al. Supervised contrastive learning[J]. Advances in Neural Information Processing Systems,2020,33:18661-18673.

[97] Aberdam A, Litman R, Tsiper S, et al. Sequence-to-sequence contrastive learning for text recognition[C]//Proceedings of the IEEE/CVF Conference on Computer Vision and Pattern Recognition. 2021:15302-15312.

[98] Zhang X, Zhu B, Yao X, et al. Context-based contrastive learning for scene text recognition[C]// Proceedings of the AAAI Conference on Artificial Intelligence. 2022,36(3):3353-3361.

[99] Liu H, Wang B, Bao Z, et al. Perceiving stroke-semantic context: Hierarchical contrastive learning for robust scene text recognition [C]//Proceedings of the AAAI Conference on Artificial Intelligence. 2022,36(2):1702-1710.

[100] Zhang Y, Wang Z, Zhuang J. Efficient license plate recognition via holistic position attention [C]//Proceedings of the AAAI Conference on Artificial Intelligence. 2021,35(4):3438-3446.

[101] Ronneberger O, Fischer P, Brox T. U-net: Convolutional networks for biomedical image segmentation [C]//Medical Image Computing and Computer-assisted Intervention-MICCAI 2015:18th International Conference, Munich, Germany, October 5-9, 2015, Proceedings, Part III 18. Springer International Publishing. 2015:234-241.

[102] Fan X, Zhao W. Improving robustness of license plates automatic recognition in natural scenes [J]. IEEE Transactions on Intelligent Transportation Systems,2022,23(10):18845-18854.

[103] Ke X, Zeng G, Guo W. An ultra-fast automatic license plate recognition approach for unconstrained scenarios[J]. IEEE Transactions on Intelligent Transportation Systems, 2023, 24(5):5172-5185.

[104] Shi B, Bai X, Yao C. An end-to-end trainable neural network for image-based sequence recognition and its application to scene text recognition [J]. IEEE Transactions on Pattern Analysis and Machine Intelligence,2016,39(11):2298-2304.

[105] Wu C, Xu S, Song G, et al. How many labeled license plates are needed? [C]//Pattern Recognition and Computer Vision: First Chinese Conference, PRCV 2018, Guangzhou, China, November 23-26, 2018, Proceedings, Part IV 1. Springer International Publishing. 2018: 334-346.

[106] Chen C L P, Wang B. Random-positioned license plate recognition using hybrid broad learning system and convolutional networks[J]. IEEE Transactions on Intelligent Transportation Systems, 2020,23(1):444-456.

[107] Chen S L, Liu Q, Chen F, et al. End-to-end multi-line license plate recognition with cascaded perception[C]//International Conference on Document Analysis and Recognition. Cham: Springer Nature Switzerland. 2023:274-289.

[108] Gonçalves G R, Diniz M A, Laroca R, et al. Real-time automatic license plate recognition through deep multi-task networks[C]//2018 31st SIBGRAPI Conference on Graphics, Patterns and Images(SIBGRAPI). IEEE. 2018:110-117.

[109] Kingma D P. Adam: A method for stochastic optimization[C]//Proceedings of the International Conference on Learning Representations(ICLR). 2015.

[110] Li C, Liu W, Guo R, et al. PP-OCRv3: More attempts for the improvement of ultra lightweight OCR system[J]. arXiv preprint arXiv:2206.03001, 2022.

[111] Masood S Z, Shu G, Dehghan A, et al. License plate detection and recognition using deeply learned convolutional neural networks[J]. arXiv preprint arXiv:1703.07330, 2017.

[112] Sun M, Zhou F, Yang C, et al. Image generation framework for unbalanced license plate data set[C]//2019 International Conference on Data Mining Workshops(ICDMW). IEEE. 2019:883-889.

[113] Wang B, Xiao H, Zheng J, et al. Character segmentation and recognition of variable-length license plates using ROI detection and broad learning system[J]. Remote Sensing, 2022, 14(7):1560.

[114] Wang D, Tian Y, Geng W, et al. LPR-Net: Recognizing Chinese license plate in complex environments[J]. Pattern Recognition Letters, 2020, 130:148-156.

[115] Wang T, Wang W, Li C, et al. Efficient license plate recognition via parallel position-aware attention[C]//Chinese Conference on Pattern Recognition and Computer Vision(PRCV). Cham: Springer Nature Switzerland. 2022:346-360.

[116] Zhang K, Zhang Z, Li Z, et al. Joint face detection and alignment using multitask cascaded convolutional networks[J]. IEEE Signal Processing Letters, 2016, 23(10):1499-1503.

[117] Jain V, Sasindran Z, Rajagopal A, et al. Deep automatic license plate recognition system[C]//Proceedings of the Tenth Indian Conference on Computer Vision, Graphics and Image Processing. 2016:1-8.

[118] Qiao L, Chen Y, Cheng Z, et al. Mango: A mask attention guided one-stage scene text spotter[C]//Proceedings of the AAAI Conference on Artificial Intelligence. 2021, 35(3):2467-2476.

[119] Qin S, Liu S. Towards end-to-end car license plate location and recognition in unconstrained scenarios[J]. Neural Computing and Applications, 2022, 34(24):21551-21566.

[120] Wang Y, Bian Z P, Zhou Y, et al. Rethinking and designing a high-performing automatic license plate recognition approach[J]. IEEE Transactions on Intelligent Transportation Systems, 2021, 23(7):8868-8880.

[121] Yang Y, Xi W, Zhu C, et al. HomoNet: Unified license plate detection and recognition in complex scenes[C]//International Conference on Collaborative Computing: Networking, Applications and Worksharing. Cham: Springer International Publishing. 2020:268-282.

[122] Zheng Z, Wang P, Liu W, et al. Distance-IoU loss: Faster and better learning for bounding box regression[C]//Proceedings of the AAAI Conference on Artificial Intelligence. 2020, 34(7):

12993-13000.

[123] Zhuang J, Hou S, Wang Z, et al. Towards human-level license plate recognition[C]//Proceedings of the European Conference on Computer Vision(ECCV). 2018:306-321.

[124] Atienza R. Data augmentation for scene text recognition[C]//Proceedings of the IEEE/CVF International Conference on Computer Vision. 2021:1561-1570.

[125] Bochkovskiy A, Wang C Y, Liao H Y M. Yolov4:Optimal speed and accuracy of object detection [J]. arXiv preprint arXiv:2004.10934,2020.

[126] Cao Y, Fu H, Ma H. An end-to-end neural network for multi-line license plate recognition[C]// 2018 24th International Conference on Pattern Recognition(ICPR). IEEE. 2018:3698-3703.

[127] Chen S L, Tian S, Liu Q, et al. Vertex adjustment loss for multidirectional license plate detection and recognition[C]//2022 IEEE Smartworld, Ubiquitous Intelligence & Computing, Scalable Computing & Communications, Digital Twin, Privacy Computing, Metaverse, Autonomous & Trusted Vehicles (SmartWorld/UIC/ScalCom/DigitalTwin/PriComp/Meta). IEEE. 2022: 285-292.

[128] Cheng Z, Bai F, Xu Y, et al. Focusing attention: Towards accurate text recognition in natural images[C]//Proceedings of the IEEE International Conference on Computer Vision. 2017: 5076-5084.

[129] Datondji S R E, Dupuis Y, Subirats P, et al. A survey of vision-based traffic monitoring of road intersections[J]. IEEE Transactions on Intelligent Transportation Systems, 2016, 17(10): 2681-2698.

[130] He K, Gkioxari G, Dollár P, et al. Mask R-CNN[C]//Proceedings of the IEEE International Conference on Computer Vision. 2017:2961-2969.

[131] Henry C, Ahn S Y, Lee S W. Multinational license plate recognition using generalized character sequence detection[J]. IEEE Access,2020,8:35185-35199.

[132] Huang Q, Cai Z, Lan T. A single neural network for mixed style license plate detection and recognition[J]. IEEE Access,2021,9:21777-21785.

[133] Li Z J, Chen S L, Liu Q, et al. Anchor-free location refinement network for small license plate detection[C]//Chinese Conference on Pattern Recognition and Computer Vision (PRCV). Cham:Springer Nature Switzerland. 2022:506-519.

[134] Liu S, Qi L, Qin H, et al. Path aggregation network for instance segmentation[C]//Proceedings of the IEEE Conference on Computer Vision and Pattern Recognition. 2018:8759-8768.

[135] Lu Q, Liu Y, Huang J, et al. License plate detection and recognition using hierarchical feature layers from CNN[J]. Multimedia Tools and Applications,2019,78:15665-15680.

[136] Meng A, Yang W, Xu Z, et al. A robust and efficient method for license plate recognition[C]// 2018 24th International Conference on Pattern Recognition(ICPR). IEEE. 2018:1713-1718.

[137] Paidi V, Fleyeh H, Håkansson J, et al. Smart parking sensors, technologies and applications for open parking lots:A review[J]. IET Intelligent Transport Systems,2018,12(8):735-741.

[138] Selmi Z, Halima M B, Alimi A M. Deep learning system for automatic license plate detection and recognition [C]//2017 14th IAPR International Conference on Document Analysis and

Recognition(ICDAR). IEEE. 2017,1:1132-1138.
[139] Shi B, Wang X, Lyu P, et al. Robust scene text recognition with automatic rectification[C]// Proceedings of the IEEE Conference on Computer Vision and Pattern Recognition. 2016: 4168-4176.
[140] Wang W, Yang J, Chen M, et al. A light CNN for end-to-end car license plates detection and recognition[J]. IEEE Access, 2019, 7:173875-173883.
[141] Zhou X, Cheng Y, Jiang L, et al. FAFEnet: A fast and accurate model for automatic license plate detection and recognition[J]. IET Image Processing, 2023, 17(3):807-818.